Physical Science

Reading Comprehension Guide

HOLT, RINEHART AND WINSTON

A Harcourt Education Company

Orlando • Austin • New York • San Diego • Toronto • London

TO THE STUDENT

Do you need to review the concepts in the text? If so, this booklet will help you. *The Reading Comprehension Guide* is an important tool to help you organize what you have learned from the chapter so that you can succeed in your studies. The booklet contains a Directed Reading worksheet as well as a Vocabulary and Section Summary for each chapter.

Use these worksheets in the following ways:

- as a reading guide to identify and study the main concepts of each chapter before or after you read the text
- as a place to record and review the main concepts and definitions from the text
- as a reference to determine which topics you have learned well and which topics you may need to study further

Printed in the United States of America

ISBN 13: 978-0-03-46239-9 ISBN 10: 0-03-046239-8
11 0982 12 11 10
4500243021

Contents

The World of Physical Science
Directed Reading Worksheets 1
Vocabulary and Section Summary 15

The Properties of Matter
Directed Reading Worksheets 19
Vocabulary and Section Summary 28

States of Matter
Directed Reading Worksheets 32
Vocabulary and Section Summary 39

Elements, Compounds, and Mixtures
Directed Reading Worksheets 43
Vocabulary and Section Summary 51

Matter in Motion
Directed Reading Worksheets 55
Vocabulary and Section Summary 68

Forces and Motion
Directed Reading Worksheets 72
Vocabulary and Section Summary 81

Forces in Fluids
Directed Reading Worksheets 84
Vocabulary and Section Summary 92

Work and Machines
Directed Reading Worksheets 95
Vocabulary and Section Summary 104

Energy and Energy Resources
Directed Reading Worksheets 108
Vocabulary and Section Summary 119

Heat and Heat Technology
Directed Reading Worksheets 123
Vocabulary and Section Summary 137

Introduction to Atoms
Directed Reading Worksheets 142
Vocabulary and Section Summary 146

The Periodic Table
Directed Reading Worksheets 149
Vocabulary and Section Summary 156

Chemical Bonding
Directed Reading Worksheets 158
Vocabulary and Section Summary 166

Chemical Reactions
Directed Reading Worksheets 169
Vocabulary and Section Summary 179

Chemical Compounds
Directed Reading Worksheets 184
Vocabulary and Section Summary 196

Atomic Energy
Directed Reading Worksheets 201
Vocabulary and Section Summary 208

Introduction to Electricity
Directed Reading Worksheets 210
Vocabulary and Section Summary 223

Electromagnetism
Directed Reading Worksheets 229
Vocabulary and Section Summary 239

Electronic Technology
Directed Reading Worksheets 242
Vocabulary and Section Summary 251

The Energy of Waves
Directed Reading Worksheets 256
Vocabulary and Section Summary 262

The Nature of Sound

Directed Reading Worksheets **266**
Vocabulary and Section Summary. **276**

The Nature of Light

Directed Reading Worksheets **281**
Vocabulary and Section Summary. **297**

Light and Our World

Directed Reading Worksheets **302**
Vocabulary and Section Summary. **313**

Name ______________________ Class ______________ Date ______________

Skills Worksheet

Directed Reading A

Section: Exploring Physical Science

______ **1.** What two activities are basic to most of science?
- **a.** getting good grades and asking questions
- **b.** observing the world and asking questions
- **c.** paying attention in class and memorizing
- **d.** observing science and reading the answers

THAT'S SCIENCE!

2. Science starts with gathering ______________________ about the natural world.

3. The knowledge obtained by observation and the testing of laws and principles is called ______________________.

WHAT IS PHYSICAL SCIENCE?

______ **4.** What is physical science?
- **a.** the study of scientific methods
- **b.** the study of knowledge
- **c.** the study of nonliving matter
- **d.** the study of living things

5. The stuff that everything is made of is ______________________.

6. The ability to do work is called ______________________.

7. All moving objects have energy of ______________________.

8. All matter, including matter that isn't moving, has ______________________.

BRANCHES OF PHYSICAL SCIENCE

9. The two major branches of physical science are ______________________ and ______________________.

10. The study of substances made of carbon is called ______________________.

11. The study of all forms of matter, including how matter interacts with other matter, is called ______________________.

12. An important part of chemistry is the study of the structure and properties of ______________________.

13. Changes in substances, called ______________________, take place around us all of the time.

14. List three subjects included in the science of chemistry.

____

____

____

15. An important concern of physics is the way that ______________________ affects matter.

16. The examination of different forms of energy is part of the study of

______________________.

17. List some of the things that are parts of physics.

____

____

____

____

____

____

18. Compasses help us find our way because of the existence of

______________________.

PHYSICAL SCIENCE: ALL AROUND YOU

19. A person who studies the atmosphere is called a(n)

______________________.

20. Do meteorologists need to have a knowledge of physical science? Explain your answer.

____

____

____

21. The study of the origin, history, and structure of Earth is called

______________________.

22. A person who studies the chemistry of rocks, minerals, and soil is a(n)

______________________.

23. Do geochemists need to have a knowledge of physical science? Explain your answer.

____

____

____

24. A knowledge of physical science will help a biologist understand how animals

get ______________________ from food.

25. How are life science and physical science related? Explain your answer.

____

____

____

Name ______________________ Class ______________ Date ____________

Skills Worksheet

Directed Reading A

Section: Scientific Methods

WHAT ARE SCIENTIFIC METHODS?

______ **1.** What is the series of steps scientists use to answer questions and solve problems?
a. observations
b. formulations
c. flowcharts
d. scientific methods

2. List the steps that are considered scientific methods.

____

____

____

____

____

____

ASKING A QUESTION

______ **3.** What does asking questions help scientists to do?
a. find answers with less investigation
b. focus the purpose of an investigation
c. ask questions and memorize answers
d. know where to look up the answers

4. Any use of the senses to gather information is called

______________________.

5. Observations made with tools are called ______________________.

6. Efficiency compares energy output with ______________________.

7. Explain why the efficiency of a boat is important.

____

____

____

8. What real world question did the two engineers James Czarnowski and Michael Triantafyllou explore?

____

____

FORMING A HYPOTHESIS

______ **9.** An explanation that is based on observation and that can be tested is
a. an observation.
b. a hypothesis.
c. efficiency.
d. a conclusion.

______ **10.** After a scientist has asked questions and made observations, she is ready to
a. answer the questions.
b. explain the answers.
c. start a different investigation.
d. form a hypothesis.

11. How are observations related to the process of forming a hypothesis?

____

____

____

12. A good hypothesis should be ______________.

13. What is the problem with a hypothesis that can't be tested? Explain your answer.

____

____

____

14. What was the hypothesis that Czarnowski formed?

____

____

____

15. How did Czarnowski form his hypothesis? Explain your answer.

____

____

____

16. A good way to make a prediction about a hypothesis is by stating it

in a(n) ________________ format.

17. How might Czarnowski and Triantafyllou have stated their prediction in an if-then format?

____

____

TESTING THE HYPOTHESIS

______**18.** Testing a hypothesis helps you determine if the hypothesis is

a. a reasonable answer to your question.
b. a controlled experiment.
c. efficient.
d. an adaptation.

______**19.** If your tests show that your hypothesis is way off the mark, you will want to

a. change the topic you are studying.
b. buy new measurement tools.
c. repeat the tests until you get the results you want.
d. repeat the tests, then change the hypothesis if necessary.

______**20.** A controlled experiment compares results from experimental groups with

a. results from other experimental groups.
b. results from other investigations.
c. results from a control group.
d. results from past experiments.

21. The purpose of a controlled experiment is to ____________________ a hypothesis.

22. In a controlled experiment, the control group and the experimental groups are the same except for a factor in the experimental groups called a(n)

____________________.

23. Is a controlled experiment always possible? Explain your answer.

____

____

____

24. How did Czarnowski and Triantafyllou decide to test their hypothesis?

____

____

25. Pieces of information gathered through observation or experimentation are called ______________________.

26. What three kinds of the data were collected during the *Proteus* experiment?

____

____

____

ANALYZING THE RESULTS

27. After you run an experiment and collect data, you must ______________________ the data to see if the results support your hypothesis.

28. Organizing data into ______________________ and ______________________ can make information easier to use.

29. What kind of graph can be used to make a comparison?

____

DRAWING CONCLUSIONS

______**30.** What must you do at the end of an experiment?

a. Draw a conclusion.
b. Analyze a graph.
c. Draw a picture.
d. Analyze a chart.

31. Give examples of conclusions you might draw after an investigation.

____

____

____

32. What did the two engineers conclude after the trials of the *Proteus*?

____

____

____

COMMUNICATING RESULTS

33. What are some ways to communicate results of a scientific investigation?

____

____

____

34. Why is it important to communicate results of a scientific investigation?

____

____

____

Name ______________________ Class ______________ Date ____________

Skills Worksheet

Directed Reading A

Section: Scientific Models

1. What can scientists build to test their hypotheses?

____

MODELS IN SCIENCE

______ **2.** What is a pattern, plan, representation, or description designed to show the structure or workings of an object, system, or concept called?

a. a test
b. a model
c. a hypothesis
d. a *Proteus*

3. Scientists use models to represent ____________________ or systems.

4. A model uses something ____________________ to help us understand something that is not familiar.

5. List the three common types of scientific models.

____

____

____

6. What can be helpful to show things that are too small to see or how something works?

____

7. List two examples of physical models.

____

____

8. A mathematical model is made up of mathematical equations and

____________________.

9. Complex mathematical models can have many ____________________.

10. Omitting a variable in a complex mathematical model can cause the model to ____________________.

11. What kind of models are based on systems of ideas or comparisons with familiar things?

____

12. How does a conceptual model make use of familiar things?

____

Match the correct description with the correct term. Write the letter in the space provided.

______ **13.** a model used to predict the weather

______ **14.** the big bang theory, which describes how the planets and galaxies were formed

______ **15.** a model of a molecule

a. conceptual model

b. physical model

c. mathematical model

16. A model can help you see things in your mind that are very

______________ and very ______________.

17. Give examples of how models help us picture things that are very difficult to see.

____

____

____

USING MODELS TO BUILD SCIENTIFIC KNOWLEDGE

______ **18.** An explanation for many hypotheses and observations is called a

a. model.
b. law.
c. variable.
d. theory.

19. What are theories based on?

____

____

____

20. A theory explains observations you've made and helps you

______________ what will happen in other tests.

21. Explain how scientists use models to test theories.

____

____

____

22. What is a summary of many experimental results and observations called?

____

23. A scientific law is formed only after many experimental results and

_______________.

24. What do laws describe?

____

25. How do theories and laws differ?

____

____

26. What law says that the total mass of materials formed is the same as the total mass of the starting materials?

____

Name ________________________ Class ______________ Date ____________

Skills Worksheet

Directed Reading A

Section: Tools, Measurement, and Safety

______ **1.** What is a *tool*?

a. something that helps you do a task
b. hardware that helps you do a task
c. anything that helps you build something
d. anything with a handle

TOOLS IN SCIENCE

2. One way that tools are used in scientific investigations is for collecting ______________________ by taking measurements.

3. List four examples of tools used for measuring.

____

____

____

____

4. List three examples of tools that help you analyze or communicate data.

____

____

____

5. List two examples of units of measure used in the distant past.

____

____

6. Scientists all over the world use the ______________________, or the metric system, in their work.

7. What does "SI" stand for?

____

8. All SI units are based on multiples of ______________________, so changing from one unit to another is easy.

MAKING MEASUREMENTS

Match the correct description with the correct term. Write the letter in the space provided.

______ **9.** the measure of the size of an object in cubic meters or liters

______ **10.** the amount of matter in a given volume; the ratio of the mass to the volume in an object

______ **11.** the measure of how long something is in millimeters and meters

______ **12.** the measure of how hot or cold something is in degrees Celsius

______ **13.** the amount of matter in an object in grams, kilograms, or metric tons

a. mass

b. temperature

c. length

d. volume

e. density

14. What is the SI unit for length?

____

15. What is the amount of matter in an object?

____

16. What is the SI unit for volume?

____

17. Describe how the liter is related to the meter.

____

____

18. How are volume and density related?

____

____

19. What is the SI unit for temperature?

____

20. What units of temperature do scientists frequently use?

____

SAFETY RULES

21. List five basic safety rules to follow when doing scientific experiments.

____

____

____

____

____

Match the labels to the icons. Write the letters in the spaces provided.

______ **22.** heating safety — **a.**

______ **23.** eye protection — **b.**

______ **24.** sharp object — **c.**

______ **25.** electric safety — **d.**

______ **26.** clothing protection — **e.**

______ **27.** animal safety — **f.**

______ **28.** hand safety — **g.**

______ **29.** chemical safety — **h.**

______ **30.** plant safety — **i.**

Name ______________________ Class ______________ Date ____________

Skills Worksheet

Vocabulary and Section Summary

Exploring Physical Science

VOCABULARY

In your own words, write a definition of the following terms in the space provided.

1. science

____

____

2. physical science

____

____

SECTION SUMMARY

Read the following section summary.

- Science is a process of gathering knowledge about the natural world.
- Physical science is the study of matter and energy.
- Physical science is divided into the study of physics and chemistry.
- Chemistry studies the structure and properties of matter and how matter changes.
- Physics looks at energy and the way that energy affects matter.
- A knowledge of physical science is important for many areas of science, such as geology and biology.

Name ______________________ Class ______________ Date ____________

Skills Worksheet

Vocabulary and Section Summary

Scientific Methods

VOCABULARY

In your own words, write a definition of the following terms in the space provided.

1. scientific methods

____

____

2. observation

____

____

3. hypothesis

____

____

4. data

____

____

SECTION SUMMARY

Read the following section summary.

- Scientific methods are the ways in which scientists answer questions and solve problems.
- Asking a question usually results from making an observation. Questioning is often the first step of using scientific methods.
- A hypothesis is a possible explanation or answer to a question. A good hypothesis is testable.
- After testing a hypothesis, you should analyze your results. Analyzing is usually done by using calculations, tables, and graphs.
- After analyzing your results, you should draw conclusions about whether your hypothesis is supported.
- Communicating your results allows others to check or continue your work. You can communicate through reports, posters, and the Internet.

Name ______________________ Class ______________ Date ____________

Skills Worksheet

Vocabulary and Section Summary

Scientific Models

VOCABULARY

In your own words, write a definition of the following terms in the space provided.

1. model

____

____

2. theory

____

____

3. law

____

____

SECTION SUMMARY

Read the following section summary.

- A model uses familiar things to describe unfamiliar things.
- Physical, mathematical, and conceptual models are commonly used in science.
- A scientific theory is an explanation for many hypotheses and observations.
- A scientific law summarizes experimental results and observations. It describes what happens but not why.

Name ______________________ Class ______________ Date ______________

Skills Worksheet

Vocabulary and Section Summary

Tools, Measurement, and Safety

VOCABULARY

In your own words, write a definition of the following terms in the space provided.

1. mass

____

____

2. volume

____

____

3. density

____

____

4. temperature

____

____

SECTION SUMMARY

Read the following section summary.

- Tools are used to make observations, take measurements, and analyze data.
- The International System of Units (SI) is the standard system of measurement.
- Length, volume, mass, and temperature are types of measurement.
- Density is the amount of matter in a given volume.
- Safety symbols are for your protection.

Name ______________________ Class ____________ Date ____________

Skills Worksheet

Directed Reading A

Section: What Is Matter?

MATTER

1. What characteristic do a human, hot soup, the metal wires in a toaster, and the glowing gases in a neon sign have in common?

____

2. What is matter?

____

MATTER AND VOLUME

_____ **3.** What unit would you use to measure the amount of water in a lake?

a. grams (g)
b. liters (L)
c. meters (m)
d. milliliters (mL)

_____ **4.** What unit would you use to measure the volume of soda in a can?

a. centimeters (cm)
b. grams (g)
c. liters (L)
d. milliliters (mL)

5. What is volume?

____

6. Things with ______________ cannot share the same space at the same time.

7. To measure a volume of water in a graduated cylinder, you should look at the bottom of the curve at the surface of the water called the ______________.

8. The volume of solid objects is commonly expressed in ______________ units.

9. What three dimensions are needed to find the volume of a rectangular solid?

____

____

____

10. How could the volume of a 12-sided object be found using water and a graduated cylinder?

____

11. Why can you express the volume of the 12-sided object measured by this method in cubic units?

____

MATTER AND MASS

_____ **12.** The amount of matter in an object is its
- **a.** volume.
- **b.** length.
- **c.** meniscus.
- **d.** mass.

_____ **13.** The SI unit of mass is the
- **a.** newton.
- **b.** liter.
- **c.** kilogram.
- **d.** pound.

_____ **14.** The SI unit of weight is the
- **a.** newton.
- **b.** liter.
- **c.** kilogram.
- **d.** pound.

_____ **15.** One newton is equal to the weight of an object that has
- **a.** a mass of 100 g on the moon.
- **b.** a volume of 1 m^3 on Earth.
- **c.** a mass of 1,000 g on Earth.
- **d.** a mass of 100 g on Earth.

16. What is the only way to change the mass of an object?

____

____

For each description, write whether it applies to mass or to weight.

______________ **17.** is always constant no matter where the object is located

______________ **18.** is a measure of the gravitational force on an object

______________ **19.** is measured using a spring scale

______________ **20.** is expressed in grams (g), kilograms (kg), or milligrams (mg)

______________ **21.** is expressed in newtons (N)

______________ **22.** is less on the moon than on Earth

______________ **23.** is a measure of the amount of matter in the object

INERTIA

_____ **24.** The tendency of an object to resist a change in motion is known as
- **a.** mass.
- **b.** gravitation.
- **c.** inertia.
- **d.** weight.

25. What is needed in order to cause an object at rest to move, or an object in motion to change its direction or speed?

____

____

26. How does mass affect the inertia of an object?

____

____

____

27. Why is it harder to get a cart full of potatoes moving than one that is empty?

____

____

____

____

Name ______________________ Class ______________ Date ____________

Skills Worksheet

Directed Reading A

Section: Physical Properties

PHYSICAL PROPERTIES

______ **1.** A characteristic of matter that can be observed or measured without changing the identity of the matter is a

a. matter property.
b. physical property.
c. chemical property.
d. volume property.

______ **2.** Some examples of physical properties are

a. color, odor, and age.
b. color, odor, and speed.
c. color, odor, and magnetism.
d. color, odor, and anger.

Match the correct example with the correct physical property. Write the letter in the space provided.

______ **3.** Aluminum can be flattened into sheets of foil.

______ **4.** An ice cube floats in a glass of water.

______ **5.** Copper can be pulled into thin wires.

______ **6.** Plastic foam protects you from hot liquid.

______ **7.** Flavored drink mix dissolves in water.

______ **8.** An onion gives off a very distinctive smell.

______ **9.** A golf ball has more mass than a table tennis ball.

a. state
b. solubility
c. thermal conductivity
d. malleability
e. odor
f. ductility
g. density

10. Density is the ______________________ that describes the relationship between mass and volume.

11. Objects such as a cotton ball and a small tomato can occupy similar volumes but vary greatly in ______________________.

12. If you pour different liquids into a graduated cylinder, the liquids will form layers based upon differences in the ______________________ of each liquid.

13. Which layer of liquid would settle on the bottom of a graduated cylinder?

____

14. Where will the least dense liquid be found?

____

15. Why would 1 kg of lead be less awkward to carry around than 1 kg of feathers?

____

____

____

16. What will happen to a solid object made from matter with a greater density than water when it is dropped into water?

____

____

____

17. How will knowing the density of a substance help you determine whether an object made from that material will float in water?

____

____

____

18. What is the equation for density?

____

19. What do D, V, and m stand for in the equation for density?

____

____

____

20. The units for density take the form of a mass unit divided by a(n)

______________________ unit.

21. What are two reasons why density is a useful property for identifying substances?

____

____

PHYSICAL CHANGES DO NOT FORM NEW SUBSTANCES

22. A change that affects only the physical properties of a substance is known as a(n) ____________________.

23. What kind of changes are melting and freezing?

____

Identify which of the following activities represent physical changes by writing PC in the space provided if they cause only physical changes. Put an X beside any that do not.

______**24.** sanding a piece of wood

______**25.** baking bread

______**26.** crushing an aluminum can

______**27.** melting an ice cube

______**28.** dissolving sugar in water

______**29.** molding a piece of silver

30. When a substance undergoes a physical change, its ____________________ does not change.

31. What is changed when matter undergoes a physical change? Give an example to explain your answer.

____

____

____

____

____

____

____

Name ______________________________ Class ________________ Date ____________

Skills Worksheet

Directed Reading A

Section: Chemical Properties

CHEMICAL PROPERTIES

______ **1.** The property of matter that describes its ability to change into new matter with different properties is known as a
a. chemical change.
b. physical change.
c. chemical property.
d. physical property.

______ **2.** The chemical property that describes the ability of two or more substances to combine to form new substances is called
a. reactivity.
b. flammability.
c. density.
d. solubility.

______ **3.** The ability of a substance to burn is a chemical property known as
a. reactivity.
b. flammability.
c. density.
d. solubility.

______ **4.** An iron nail is reactive with
a. rubbing alcohol.
b. other iron nails.
c. wood in a house.
d. oxygen in the air.

______ **5.** Which of the following statements is true about characteristic properties of matter?
a. Characteristic properties depend on the size of the sample.
b. Characteristic properties may be either physical or chemical properties.
c. Characteristic properties involve only chemical properties.
d. Characteristic properties involve only the physical nature of the matter.

6. Describe the ways that burning changes the nature of wood.

____

____

7. A substance always has ______________________ properties, even though they are difficult to observe.

8. Scientists use ______________________ properties to help them identify and classify matter.

CHEMICAL CHANGES AND NEW SUBSTANCES

______ **9.** Chemical changes are the processes by which substances
a. move from place to place.
b. change into new substances.
c. change in their physical properties.
d. become greater in mass.

______ **10.** Which of the following would NOT be considered an example of a chemical change?
a. the bubbling action of effervescent tablets
b. the green coating on copper statues
c. the melting of a Popsicle
d. the burning of rocket fuel

11. How do you know that baking a cake involves chemical changes?

____

____

____

____

12. List some signs or clues that show that a change you are observing is a chemical change.

____

____

____

____

13. Because _______________ change the identity of the substances involved, they are hard to reverse.

14. How could some chemical changes be reversed? Give an example.

____

____

____

____

PHYSICAL VERSUS CHEMICAL CHANGES

______ **15.** What is the most important question to ask to determine whether a change is physical or chemical?
- **a.** Was there a color change?
- **b.** Did the composition change?
- **c.** Was there a change in size?
- **d.** Did the change involve a change in state?

______ **16.** What is the name of the process by which water is broken down into hydrogen and oxygen using an electric current?
- **a.** electrolysis
- **b.** decomposition
- **c.** reactivity
- **d.** reversibility

17. During ______________________, the composition of a substance does not change.

Identify whether the following changes are physical changes or chemical changes. Label each change either PC for physical change or CC for chemical change.

______ **18.** mixing vinegar and baking soda

______ **19.** grinding baking soda into a powder

______ **20.** souring milk

______ **21.** melting an ice cream bar

______ **22.** burning a wooden match

______ **23.** shooting off fireworks

______ **24.** mixing drink mix into water

______ **25.** bending an iron nail

Name ______________________ Class ______________ Date ______________

Skills Worksheet

Vocabulary and Section Summary

What Is Matter?

VOCABULARY

In your own words, write a definition of the following terms in the space provided.

1. matter

____

____

2. volume

____

____

3. meniscus

____

____

4. mass

____

____

5. weight

____

____

6. inertia

____

____

SECTION SUMMARY

Read the following section summary.

- Two properties of matter are volume and mass.
- Volume is the amount of space taken up by an object.
- The SI unit of volume is the liter (L).
- Mass is the amount of matter in an object.
- The SI unit of mass is the kilogram (kg).
- Weight is a measure of the gravitational force on an object, usually in relation to the Earth.
- Inertia is the tendency of an object to resist being moved or, if the object is moving, to resist a change in speed or direction. The more massive an object is, the greater its inertia.

Name ______________________ Class ______________ Date ____________

Skills Worksheet

Vocabulary and Section Summary

Physical Properties

VOCABULARY

In your own words, write a definition of the following terms in the space provided.

1. physical property

____

____

2. density

____

____

3. physical change

____

____

SECTION SUMMARY

Read the following section summary.

- Physical properties of matter can be observed without changing the identity of the matter.
- Examples of physical properties are conductivity, state, malleability, ductility, solubility, and density.
- Density is the amount of matter in a given space.
- Density is used to identify substances because the density of a substance is always the same at a given pressure and temperature.
- When a substance undergoes a physical change, its identity stays the same.
- Examples of physical changes are freezing, cutting, bending, dissolving, and melting.

Name ______________________ Class ______________ Date ______________

Skills Worksheet

Vocabulary and Section Summary

Chemical Properties

VOCABULARY

In your own words, write a definition of the following terms in the space provided.

1. chemical property

____

____

2. chemical change

____

____

SECTION SUMMARY

Read the following section summary.

- Chemical properties describe a substance based on its ability to change into a new substance that has different properties.
- Chemical properties can be observed only when a chemical change might happen.
- Examples of chemical properties are flammability and reactivity.
- New substances form as a result of a chemical change.
- Unlike a chemical change, a physical change does not alter the identity of a substance.

Name ______________________ Class ______________ Date ____________

Skills Worksheet

Directed Reading A

Section: Three States of Matter

1. What are the three most familiar states of matter?

____

____

____

2. What is a state of matter?

____

____

PARTICLES OF MATTER

3. Matter is made up of ____________________ and ____________________.

Match the correct description with the correct state of matter. Write the letter in the space provided.

_____	**4.** Particles do not move fast enough to overcome the strong attraction between them.	**a.** solid
_____	**5.** Particles move independently of each other.	**b.** liquid
_____	**6.** Particles are close together but can slide past one another.	**c.** gas
_____	**7.** Particles are close together and vibrate in place.	
_____	**8.** Particles move fast enough to overcome nearly all of the attraction between them.	

SOLIDS

_____ **9.** The particles of matter that make up a solid

a. have a weaker attraction than those of a liquid.
b. do not move at all.
c. do not move fast enough to overcome the force of attraction.
d. move from place to place.

10. What is a solid?

____

____

11. How are the particles in a crystalline solid arranged?

____

____

12. How are the particles in an amorphous solid arranged?

____

____

LIQUIDS

13. How do the particles of a liquid make it possible to pour juice into a glass?

____

____

14. A beaker and a cylinder each contain 350 mL of juice. What does this show you about the properties of a liquid?

____

____

15. Liquids tend to form spherical droplets because of ______________ tension.

16. Water has a lower ______________ than honey.

GASES

17. What is a gas?

____

____

18. How is it possible for one tank of helium to fill 700 balloons?

____

____

____

Name ______________________ Class ______________ Date ____________

Skills Worksheet

Directed Reading A

Section: Behavior of Gases

DESCRIBING GAS BEHAVIOR

_____ **1.** What state of matter is helium?

a. solid **c.** gas
b. liquid **d.** plasma

2. A measure of how fast the particles in an object are moving

is the ____________________.

3. Why is more gas needed to fill helium balloons on a cold day?

____

____

____

4. The amount of space that an object takes up is the ______________.

5. The volume of any gas depends upon the size of

the ____________________.

6. The amount of force exerted on a given area is

called ____________________.

7. Why does the basketball have greater pressure than the beachball?

____

____

____

____

GAS BEHAVIOR LAWS

_____ **8.** Lifting a piston on a cylinder of gas shows that when the pressure of the gas

a. increases, the temperature increases.
b. decreases, the volume increases.
c. decreases, the volume decreases.
d. increases, the volume increases.

______ **9.** All of the following remain constant for Charles's law EXCEPT
   a. the type of piston.
   b. the amount of gas.
   c. the volume of the gas.
   d. the pressure.

10. The relationship between the volume and pressure of a gas is called ______________________.

11. Weather balloons are only partially inflated before they're released into the atmosphere. Why is that?

____

____

____

____

12. Putting a balloon in the freezer is one way to demonstrate ______________________.

13. The relationship between the volume and the temperature of a gas when pressure remains constant is known as ______________________.

Name ______________________ Class ______________ Date ____________

Skills Worksheet

Directed Reading A

Section: Changes of State

ENERGY AND CHANGES OF STATE

_____ **1.** Which has the most energy?

a. particles in steam
b. particles in liquid water
c. particles in ice
d. particles in freezing water

2. When a substance changes from one physical form to another, we say the substance has had a(n) ______________________.

3. List the five changes of state.

____

____

____

____

____

MELTING: SOLID TO LIQUID

4. Could you use gallium to make jewelry? Why or why not?

____

____

____

____

5. The temperature at which a substance changes from solid to liquid is the ______________________ of the substance.

6. Melting is considered a(n) ______________________ change because energy is gained by the substance as it changes state.

FREEZING: LIQUID TO SOLID

7. A substance's ______________________ is the temperature at which it changes from a liquid to a solid.

8. What happens if energy is added or removed from a glass of ice water?

____

____

____

____

9. Freezing is considered a(n) ____________________ change because energy is removed from the substance.

EVAPORATION: LIQUID TO GAS

Match the correct definition with the correct term. Write the letter in the space provided.

______ **10.** the change of a substance from a liquid to a gas

______ **11.** the change of state from a liquid to a gas when the vapor pressure equals the atmospheric pressure

______ **12.** the pressure inside the bubbles of a boiling liquid

______ **13.** the temperature at which a liquid boils

a. boiling point
b. vapor pressure
c. evaporation
d. boiling

14. As you go higher above sea level, the ____________________ decreases and the ____________________ of a substance gets lower.

CONDENSATION: GAS TO LIQUID

15. The change of state from a gas to a liquid is ____________________.

16. At a given pressure, the condensation point for a substance is the same as its ____________________.

17. For a substance to change from a gas to a liquid, particles must ____________________.

SUBLIMATION: SOLID TO GAS

18. Solid carbon dioxide isn't ice. So why is it called "dry ice"?

____

____

____

19. The change of state from a solid to a gas is called ______________.

CHANGE OF TEMPERATURE VS. CHANGE OF STATE

20. The speed of the particles in a substance changes when the

______________ changes.

21. The temperature of a substance does not change before the

______________ is complete.

Name ______________________ Class ______________ Date ____________

Skills Worksheet

Vocabulary and Section Summary

Three States of Matter

VOCABULARY

In your own words, write a definition of the following terms in the space provided.

1. states of matter

____

____

2. solid

____

____

3. liquid

____

____

4. surface tension

____

____

5. viscosity

____

____

6. gas

____

____

SECTION SUMMARY

Read the following section summary.

- The three most familiar states of matter are solid, liquid, and gas.
- All matter is made of tiny particles called atoms and molecules that attract each other and move constantly.
- A solid has a definite shape and volume.
- A liquid has a definite volume but not a definite shape.
- A gas does not have a definite shape or volume.

Name ______________________ Class ______________ Date ____________

Skills Worksheet

Vocabulary and Section Summary

Behavior of Gases

VOCABULARY

In your own words, write a definition of the following terms in the space provided.

1. temperature

____

____

2. volume

____

____

3. pressure

____

____

4. Boyle's law

____

____

5. Charles's law

____

____

SECTION SUMMARY

Read the following section summary.

- Temperature measures how fast the particles in an object are moving.
- Gas pressure increases as the number of collisions of gas particles increases.
- Boyle's law states that if the temperature doesn't change, the volume of a gas increases as the pressure decreases.
- Charles's law states that if the temperature doesn't change, the volume of a gas increases.

Name ______________________ Class ______________ Date ____________

Skills Worksheet

Vocabulary and Section Summary

Changes of State

VOCABULARY

In your own words, write a definition of the following terms in the space provided.

1. change of state

____

____

2. melting

____

____

3. evaporation

____

____

4. boiling

____

____

5. condensation

____

____

6. sublimation

____

____

SECTION SUMMARY

Read the following section summary.

- A change of state is the conversion of a substance from one physical form to another.
- Energy is added during endothermic changes. Energy is removed during exothermic changes.
- The freezing point and the melting point of a substance are the same temperature.
- Both boiling and evaporation result in a liquid changing to a gas.
- Condensation is the change of a gas to a liquid. It is the reverse of evaporation.
- Sublimation changes a solid directly to a gas.
- The temperature of a substance does not change during a change of state.

Name ______________________ Class ______________ Date ______________

Skills Worksheet

Directed Reading A

Section: Elements

______ **1.** Which of the following is NOT a physical or chemical change?
a. crushing
b. weighing
c. melting
d. passing electric current

ELEMENTS, THE SIMPLEST SUBSTANCES

2. A pure substance that cannot be separated into simpler substances by physical or chemical means is a(n) ______________________.

3. A substance that contains only one type of particle is a(n) ______________________.

PROPERTIES OF ELEMENTS

4. The amount of an element present does not affect the element's ______________________.

5. Why does a helium-filled balloon float up when it is released?

____

____

Look at each property listed below. If it is a characteristic property of elements, write CP on the line. If it is not a characteristic property, write N.

______ **6.** size
______ **7.** melting point
______ **8.** density
______ **9.** shape
______ **10.** mass
______ **11.** volume
______ **12.** color
______ **13.** hardness
______ **14.** flammability
______ **15.** weight
______ **16.** reactivity with acid

CLASSIFYING ELEMENTS BY THEIR PROPERTIES

17. What are two common properties that most terriers share?

____

____

18. All elements can be classified as metals, metalloids, or

______________________.

19. An element that is shiny and that conducts heat and electric current well is

a(n) ______________________.

20. An element that conducts heat and electric current poorly, and can be a solid,

liquid, or gas is a(n) ______________________.

21. Elements that have properties of both metals and nonmetals

are ______________________.

Indicate whether the description applies to a metal, a nonmetal, or a metalloid. Write the correct letter in the space provided.

______**22.** are malleable

______**23.** are dull or shiny

______**24.** are poor conductors

______**25.** tend to be brittle and unmalleable as solids

______**26.** are always shiny

______**27.** are also called semiconductors

______**28.** are always dull

______**29.** are somewhat ductile

______**30.** include boron, silicon, antimony

______**31.** include lead, tin, copper

______**32.** include sulfur, iodine, neon

a. metalloids

b. nonmetals

c. metals

Name ______________________ Class ______________ Date ______________

Skills Worksheet

Directed Reading A

Section: Compounds

1. List three examples of compounds you encounter every day.

____

____

COMPOUNDS: MADE OF ELEMENTS

______ **2.** Which of the following is NOT true about compounds?
- **a.** Compounds are combinations of elements that join in specific ratios according to their masses.
- **b.** The mass ratio of a specific compound is always the same.
- **c.** Compounds are random combinations of elements.
- **d.** Different mass ratios mean different compounds.

3. When two or more elements are joined by chemical bonds to form a new pure substance, we call that new substance a(n) ______________________.

4. A compound is different from the ______________________ that reacted to form it.

PROPERTIES OF COMPOUNDS

______ **5.** Which of the following statements is true about the properties of compounds?
- **a.** A property of all compounds is to react with acid.
- **b.** Each compound has its own physical properties.
- **c.** Compounds cannot be identified by their chemical properties.
- **d.** A compound has the same properties as the elements that form it.

6. Sodium and chlorine can be extremely dangerous in their elemental form. How is it possible that we can eat them in a compound?

____

____

____

Match the correct description with the correct term. Write the letter in the space provided.

_____ **7.** a poisonous, greenish yellow gas

_____ **8.** table salt

_____ **9.** a soft, silvery white metal that reacts violently with water

a. sodium chloride

b. chlorine

c. sodium

BREAKING DOWN COMPOUNDS

10. What compound helps give carbonated beverages their "fizz"?

____

____

11. Which elements make up the compound that helps give carbonated beverages their "fizz"?

____

____

12. The only way to break down a compound is through

a(n) ______________________ change.

COMPOUNDS IN YOUR WORLD

13. Aluminum is produced by breaking down the compound

______________________.

14. Plants use the compound ______________________ in photosynthesis to make carbohydrates.

Name ______________________ Class ______________ Date ____________

Skills Worksheet

Directed Reading A

Section: Mixtures

1. A pizza is a(n) ____________________.

PROPERTIES OF MIXTURES

2. A combination of two or more substances that are not chemically combined is a(n) ____________________.

3. When two or more materials combine chemically, they form a(n) ____________________.

4. How can you tell that a pizza is a mixture?

____

____

5. Mixtures are separated through ____________________ changes.

Match the correct method of separation with the each substance. Write the letter in the space provided. Each method may be used only once.

_____ **6.** crude oil

_____ **7.** a mixture of aluminum and iron

_____ **8.** parts of blood

_____ **9.** sulfur and salt

a. distillation

b. magnet

c. filter

d. centrifuge

10. Granite can be pink or black, depending on the ____________________ of feldspar, mica, and quartz.

SOLUTIONS

_____ **11.** Which of the following is NOT true of solutions?

a. They contain a dissolved substance called a solute.

b. They are composed of two or more evenly distributed substances.

c. They contain a substance called a solvent, in which another substance is dissolved.

d. They appear to be more than one substance.

12. The process in which particles of substances separate and spread evenly through a mixture is known as ____________________.

13. In a solution, the ______________________ is the substance that is dissolved, and the ______________________ is the substance in which it is dissolved.

14. Salt is ______________________ in water because it dissolves in water.

15. When two gases or two liquids form a solution, the substance that is present in the largest amount is the ______________________.

16. A solid solution of metals or nonmetals dissolved in metals is a(n) ______________________.

17. What can particles in solution NOT do because they are so small?

____

____

CONCENTRATION OF SOLUTIONS

18. A measure of the amount of solute dissolved in a solvent is called ______________________.

19. What is the difference between a dilute solution and a concentrated solution?

____

____

20. The ability of a solute to dissolve in a solvent at a certain temperature and pressure is called ______________________.

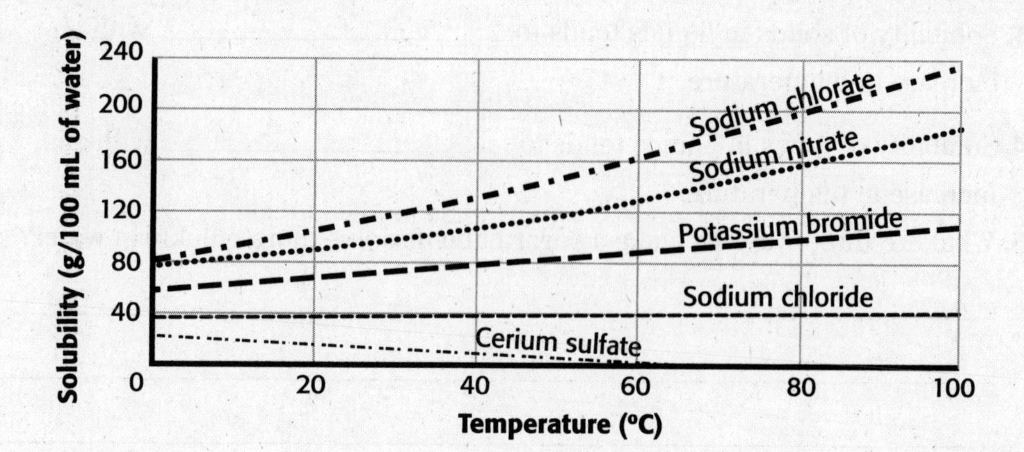

______ **21.** Look at the graph. Which solid is less soluble at higher temperatures than at lower temperatures?

a. sodium chloride
b. sodium nitrate
c. potassium bromide
d. cerium sulfate

______ **22.** Look at the graph. Which compound's solubility is least affected by temperature changes?

a. sodium chloride
b. sodium nitrate
c. potassium bromide
d. cerium sulfate

23. Solubility of solids in liquids tends to ______________________ with an increase in temperature.

24. Solubility of gases in liquids tends to ______________________ with an increase in temperature.

25. What are three ways to make a sugar cube dissolve more quickly in water?

____

____

____

SUSPENSIONS

______ **26.** Which of the following does NOT describe a suspension?

a. Particles are soluble.
b. Particles settle out over time.
c. Particles can block light.
d. Particles scatter light.

27. Why are the particles in a snow globe considered a suspension?

____

____

COLLOIDS

28. What do gelatin, milk, and stick deodorant have in common?

____

____

Match the correct description with the correct term. Write the letter in the space provided.

______ **29.** a mixture of two or more uniformly dispersed substances

______ **30.** a mixture in which particles of a material are more or less evenly dispersed throughout a liquid or gas

______ **31.** a mixture of particles that are large enough to scatter light but are not heavy enough to settle out

a. colloid
b. solution
c. suspension

Name ______________________ Class ______________ Date ____________

Skills Worksheet

Vocabulary and Section Summary

Elements

VOCABULARY

In your own words, write a definition of the following terms in the space provided.

1. element

____

____

2. pure substance

____

____

3. metal

____

____

4. nonmetal

____

____

5. metalloid

____

____

SECTION SUMMARY

Read the following section summary.

- A substance in which all of the particles are alike is a pure substance.
- An element is a pure substance that cannot be broken down into anything simpler by physical or chemical means.
- Each element has a unique set of physical and chemical properties.
- Elements are classified as metals, nonmetals, or metalloids, based on their properties.

Name ______________________ Class ______________ Date ______________

Skills Worksheet

Vocabulary and Section Summary

Compounds

VOCABULARY

In your own words, write a definition of the following term in the space provided.

1. compound

____

____

SECTION SUMMARY

Read the following section summary.

- A compound is a pure substance composed of two or more elements.
- The elements that form a compound always combine in a specific ratio according to their masses.
- Each compound has a unique set of physical and chemical properties that differ from those of the elements that make up the compound.
- Compounds can be broken down into simpler substances only by chemical changes.

Name ______________________ Class ______________ Date ____________

Skills Worksheet

Vocabulary and Section Summary

Mixtures

VOCABULARY

In your own words, write a definition of the following terms in the space provided.

1. mixture

____

____

2. solution

____

____

3. solute

____

____

4. solvent

____

____

5. concentration

____

____

6. solubility

____

____

7. suspension

____

____

8. colloid

____

____

SECTION SUMMARY

Read the following section summary.

- A mixture is a combination of two or more substances, each of which keeps its own characteristics.
- Mixtures can be separated by physical means, such as filtration and evaporation.
- A solution is a mixture that appears to be a single substance but is composed of a solute dissolved in a solvent.
- Concentration is a measure of the amount of solute dissolved in a solvent.
- The solubility of a solute is the ability of the solute to dissolve in a solvent at a certain temperature.
- Suspensions are mixtures that contain particles large enough to settle out or be filtered and to block or scatter light.
- Colloids are mixtures that contain particles that are too small to settle out or be filtered but are large enough to scatter light.

Name ______________________ Class ______________ Date ____________

Skills Worksheet

Directed Reading A

Section: Measuring Motion

1. Name something in motion that you cannot see moving.

____

____

____

OBSERVING MOTION BY USING A REFERENCE POINT

_____ **2.** An object in motion is moving in relation to an object that appears to

a. stay in place.
b. keep moving.
c. maintain constant velocity.
d. maintain constant acceleration.

_____ **3.** When an object changes position over time relative to a reference point, the object is

a. speeding.
b. accelerating.
c. decelerating.
d. moving.

4. For determining motion, the surface of Earth is a

common ______________________.

5. Why are buildings, trees, and mountains all useful reference points?

____

____

6. Can a moving object be used as a reference point? Explain.

____

____

____

SPEED DEPENDS ON DISTANCE AND TIME

7. The speed of an object depends on the distance traveled and the

______________________ taken to travel that distance.

8. The SI unit for speed is ______________________.

9. Why is it useful to calculate average speed?

____

10. Explain how to calculate average speed.

____

____

____

11. When a person drives for several hours, how does the distance traveled in one hour usually compare with the distance traveled in other hours? Explain.

____

____

12. Suppose that, on a graph showing speed, there are two lines. One line represents speed per hour, and the other line represents average speed. Will both lines be exactly alike and in the same place on the graph? Explain.

____

____

____

____

VELOCITY: DIRECTION MATTERS

13. Why wouldn't birds end up at the same destination if they are flying exactly the same speed at all times?

____

____

14. What is the difference between velocity and speed?

____

____

15. How would you calculate the resultant velocity of two velocities in the same direction?

____

____

16. How would you calculate the resultant velocity of two velocities in opposite directions? What direction is the larger velocity?

____

____

ACCELERATION

17. If your speed is not changing but your direction is changing, are you accelerating? Explain your answer.

____

____

____

18. Another name for acceleration in which velocity increases is

______________________ acceleration.

19. What are the two common terms for decrease in velocity?

____

____

20. Write the mathematical formula for calculating average acceleration.

____

____

____

21. A speedometer shows that a cyclist is going 1 m/s the 1st second, 2 m/s the 2nd second, and 3 m/s the 3rd second, as the cyclist continues straight south. How do you know the cyclist is accelerating?

____

____

____

____

22. How would acceleration be shown on a graph?

____

____

____

____

23. A graph shows a roller coaster increasing in velocity for the first eight seconds as it goes down the hill. Will the graph have an upward slope representing a roller coaster traveling down the hill? Explain your answer.

____

____

____

____

____

____

____

____

24. As long as something travels in a circle, is it always accelerating? Explain your answer.

____

____

____

____

____

____

____

Name ______________________ Class ______________ Date ____________

Skills Worksheet

Directed Reading A

Section: What Is a Force?

1. In science, a push or a pull is a(n) ____________________.

2. Any change in motion is caused by a(n) ____________________.

3. Force is expressed by a unit called the ____________________.

FORCES ACTING ON OBJECTS

4. Force always acts on a(n) ____________________.

5. Give two examples of objects on which you exert forces when you are doing your schoolwork.

____

____

6. Give one example of a force that does not cause an object to move.

____

7. What is one example of an unseen source exerting a force?

____

____

8. What is one example of an unseen receiver of a force?

____

DETERMINING NET FORCE

9. The combination of all forces acting on an object is

____________________.

10. How is net force determined if two students moving a piano exert force in the same direction?

____

____

____

____

11. Two dogs are pulling on a rope in opposite directions. The dog on the left pulls with a force of 10 N, while the dog on the right pulls with a force of 12 N. Which dog will win the tug-of-war? What is the net force?

____

____

____

____

BALANCED AND UNBALANCED FORCES

12. What will knowing the net force on an object tell you about the forces on the object?

____

____

____

____

13. When are the forces on an object *balanced?*

____

____

14. Forces are unbalanced when the net force is not equal to a certain number of newtons. What is that number?

____

15. What do you need to cause an object to start moving?

____

16. Give an example of an object that continues to move when an unbalanced force is removed.

____

____

____

____

____

Name ______________________ Class ______________ Date ____________

Skills Worksheet

Directed Reading A

Section: Friction: A Force that Opposes Motion

1. What unbalanced force causes a ball to stop rolling?

____

____

2. The force that opposes motion between two surfaces that are in contact is ______________________.

THE SOURCE OF FRICTION

3. What are two factors that affect the amount of friction between two surfaces?

____

____

4. What happens to friction if the force pushing surfaces together increases?

____

____

5. Why is more force needed to slide a large book across a table than to slide a small book across the same table?

____

____

____

6. Is the amount of friction greater between rough surfaces or smooth surfaces? Why?

____

____

____

TYPES OF FRICTION

7. What is kinetic friction?

____

____

8. What are two types of kinetic friction?

____

____

9. Which type of kinetic friction is usually greater, sliding kinetic friction or rolling kinetic friction?

____

____

____

10. What is one example of the use of sliding kinetic friction?

____

11. What is one example of the use of rolling kinetic friction?

____

12. What is static friction?

____

____

____

13. As soon as an object starts moving, what replaces static friction?

____

____

FRICTION: HARMFUL AND HELPFUL

14. What is one helpful way friction affects a car?

____

____

____

15. What is one harmful way friction affects a car?

____

____

____

16. What is a substance applied to a surface to reduce friction called?

____

17. What are three ways friction can be reduced?

____

____

____

18. What are two ways friction can be increased?

____

____

Name ______________________ Class ______________ Date ____________

Skills Worksheet

Directed Reading A

Section: Gravity: A Force of Attraction

1. Why do astronauts on the moon bounce when they walk?

____

____

2. The force of attraction between two objects that is due to their masses is

______________________.

THE EFFECTS OF GRAVITY ON MATTER

3. How can the force of gravity change the motion of an object?

____

____

____

4. Why is all matter affected by gravity?

____

____

____

5. The force that pulls you toward your pencil is the force

of ______________________.

6. Since all objects are attracted toward each other because of gravity, why can't you see the objects moving toward each other?

____

____

____

____

7. How are objects around you affected by the mass of Earth?

____

____

____

NEWTON AND THE STUDY OF GRAVITY

8. What were the two questions that Sir Isaac Newton realized were actually two parts of the same question?

____

____

____

____

____

9. What connection does legend say Newton made between the moon and a falling apple?

____

____

____

____

10. Newton summarized his ideas about gravity in a law now called

______________________.

THE LAW OF UNIVERSAL GRAVITATION

11. What is stated by the law of universal gravitation?

____

____

____

____

12. How does the law of universal gravitation explain why gravity between an elephant and Earth is greater than gravity between a cat and Earth?

____

____

____

____

____

13. How does the law of universal gravitation explain why astronauts on the moon bounce when they walk?

____

____

____

____

____

____

14. How does the gravitational force between objects that have small masses compare to the gravitational force between large objects?

____

____

____

____

____

15. Why doesn't the sun's gravitational force affect you more than Earth's gravitational force does?

____

____

____

____

16. How does the gravitational force between two objects that are close together compare to the gravitational force between two objects as they move farther apart?

____

____

____

____

____

WEIGHT AS A MEASURE OF GRAVITATIONAL FORCE

______ **17.** The measure of the gravitational force on an object is its
- **a.** mass.
- **b.** force.
- **c.** weight.
- **d.** gravity.

______ **18.** A measure of the amount of matter in an object is
- **a.** mass.
- **b.** force.
- **c.** weight.
- **d.** gravity.

______ **19.** If an object is moved from Earth to a place with greater gravitational force,
- **a.** its mass will stay the same.
- **b.** its weight will stay the same.
- **c.** its mass will increase.
- **d.** its weight will decrease.

20. On Earth, why are the words *mass* and *weight* often used to mean the same thing?

____

____

____

21. What is the SI unit of force?

____

22. Why is weight measured in newtons?

____

____

23. What is the main SI unit of mass?

____

24. Besides the kilogram, what are two units often used to measure mass?

____

____

Skills Worksheet

Vocabulary and Section Summary

Measuring Motion

VOCABULARY

In your own words, write a definition of the following terms in the space provided.

1. motion

____

____

2. speed

____

____

3. velocity

____

____

4. acceleration

____

____

SECTION SUMMARY

Read the following section summary.

- An object is in motion if it changes position over time in relation to a reference point.
- Speed is the distance traveled by an object divided by the time the object takes to travel that distance.
- Velocity is speed in a given direction.
- Acceleration is the rate at which velocity changes.
- An object can accelerate by changing speed, direction, or both.
- Speed can be represented on a graph of distance versus time.
- Acceleration can be represented by graphing velocity versus time.

Name ______________________ Class ______________ Date ____________

Skills Worksheet

Vocabulary and Section Summary

What Is a Force?

VOCABULARY

In your own words, write a definition of the following terms in the space provided.

1. force

____

____

2. newton

____

____

3. net force

____

____

SECTION SUMMARY

Read the following section summary.

- A force is a push or a pull. Forces have size and direction and are expressed in newtons.
- Force is always exerted by one object on another object.
- Net force is determined by combining forces. Forces in the same direction are added. Forces in opposite directions are subtracted.
- Balanced forces produce no change in motion. Unbalanced forces produce a change in motion.

Name ______________________ Class ______________ Date ____________

Skills Worksheet

Vocabulary and Section Summary

Friction: A Force That Opposes Motion

VOCABULARY

In your own words, write a definition of the following term in the space provided.

1. friction

____

____

SECTION SUMMARY

Read the following section summary.

- Friction is a force that opposes motion.
- Friction is caused by hills and valleys on the surfaces of two objects touching each other.
- The amount of friction depends on factors such as the roughness of the surfaces and the force pushing the surfaces together.
- Two kinds of friction are kinetic friction and static friction.
- Friction can be helpful or harmful.

Name ______________________ Class ______________ Date ____________

Skills Worksheet

Vocabulary and Section Summary

Gravity: A Force of Attraction

VOCABULARY

In your own words, write a definition of the following terms in the space provided.

1. gravity

____

____

2. weight

____

____

3. mass

____

____

SECTION SUMMARY

Read the following section summary.

- Gravity is a force of attraction between objects that is due to their masses.
- The law of universal gravitation states that all objects in the universe attract each other through gravitational force.
- Gravitational force increases as mass increases.
- Gravitational force decreases as distance increases.
- Weight and mass are not the same. Mass is the amount of matter in an object. Weight is a measure of the gravitational force on an object.

Name ______________________ Class ______________ Date ____________

Skills Worksheet

Directed Reading A

Section: Gravity and Motion

1. Suppose a baseball and a marble are dropped at the same time from the same height. Which ball would land first according to Aristotle? Explain.

____

____

GRAVITY AND FALLING OBJECTS

2. What Italian scientist argued that the mass of an object does not affect the time the object takes to fall to the ground?

____

3. Why do objects fall to the ground at the same rate?

____

4. On what two factors does acceleration depend?

____

5. Does a heavier object or a lighter object experience a greater gravitational force?

____

6. Why is a heavier object harder to accelerate than a lighter object?

____

7. Why does a heavier object fall with the same acceleration as a lighter object?

____

8. The rate at which velocity changes over time is called

____________________.

9. How is acceleration calculated?

____

____

10. At what rate do all objects accelerate toward Earth?

____

11. What equation is used to calculate the velocity (Δv) of a falling object?

____

AIR RESISTANCE AND FALLING OBJECTS

______ **12.** The force that opposes the motion of objects through air is
- **a.** gravity.
- **b.** net force.
- **c.** velocity.
- **d.** air resistance.

13. What three factors affect the amount of air resistance acting on an object?

____

____

____

14. What do you get when you subtract the force of air resistance from the force of gravity?

____

15. When a falling object stops accelerating, it has reached

________________ velocity.

16. If there were no air resistance, what would be the velocities of hailstones during a hailstorm?

____

17. The motion of a body when only the force of gravity is acting on the body is called ________________.

18. Why can free fall occur only where there is no air?

____

19. What are two places that have no air resistance?

____

____

ORBITING OBJECTS ARE IN FREE FALL

20. Is it true that an astronaut is weightless in space? Explain your answer.

____

____

21. A space shuttle follows the curve of the Earth's surface as it moves at a constant speed, and so is said to be ____________________ Earth.

22. Why don't space shuttle astronauts in orbit hit their heads on the ceiling of the falling shuttle?

____

____

____

____

23. What is centripetal force?

____

____

____

____

PROJECTILE MOTION AND GRAVITY

______ **24.** The curved path that an object follows when thrown, launched, or otherwise projected near the surface of Earth is called
- **a.** terminal velocity.
- **b.** projectile motion.
- **c.** terminal motion.
- **d.** projectile velocity.

______ **25.** The two independent components of projectile motion that combine to form a curved path are
- **a.** horizontal motion and vertical motion.
- **b.** parallel motion and vertical motion.
- **c.** horizontal motion and perpendicular motion.
- **d.** horizontal force and vertical force.

______ **26.** Motion parallel to the ground is called
- **a.** vertical motion.
- **b.** horizontal motion.
- **c.** parallel motion.
- **d.** horizontal force.

______ **27.** Everything on Earth is pulled downward toward the center by
a. acceleration.
b. projectile motion.
c. gravity.
d. vertical motion.

______ **28.** Motion perpendicular to the ground is called
a. vertical motion.
b. horizontal motion.
c. perpendicular motion.
d. perpendicular force.

______ **29.** Objects in projectile motion are pulled down by
a. acceleration.
b. horizontal motion.
c. vertical motion.
d. gravity.

______ **30.** Compared to a falling object, the downward acceleration of a thrown object is
a. the same.
b. faster.
c. slower.
d. constant.

______ **31.** If you want to hit a target with a thrown or propelled object, you must
a. aim directly at the target.
b. aim below the target.
c. aim above the target.
d. stand very close to the target.

Name ______________________ Class ______________ Date ____________

Skills Worksheet

Directed Reading A

Section: Newton's Laws of Motion

1. In 1686, what did Sir Issac Newton explain with his three laws of motion?

____

____

NEWTON'S FIRST LAW OF MOTION

2. What is Newton's first law of motion?

____

____

3. Which of Newton's laws of motion describes the motion of an object that has a net force of 0?

____

4. What are two examples of objects at rest?

____

____

5. How could an unbalanced force work on a chair at rest on the floor to make it slide across the room?

____

6. According to Newton's first law of motion, what will happen to the motion of objects moving with a certain velocity unless an unbalanced force acts on them?

____

____

7. If you were in a bumper car that stops when it hit another car, would you continue to move forward? Why or why not?

____

8. What unbalanced force acts to stop a desk that is sliding across a floor?

____

9. What does friction do to the motion of objects?

____

10. What is Newton's first law sometimes called?

____

11. What is the tendency of an object to resist being moved or, if the object is moving, to resist a change in speed or direction until an outside force acts on the object?

____

12. Why is it easier to change the motion of an object with a large mass than an object with a small mass?

____

____

____

NEWTON'S SECOND LAW OF MOTION

13. What is Newton's second law of motion?

____

____

14. What happens to the acceleration of an object as its mass decreases?

____

____

15. What happens to the acceleration of an object if the force on the object increases?

____

____

16. Why would a cart start moving faster if you gave it a hard push than if you gave it a soft push?

____

____

17. In what direction do objects accelerate?

____

18. How is the relationship of acceleration (a) to mass (m) and force (F) expressed mathematically?

____

19. Why is an apple easier to accelerate than a watermelon?

____

____

NEWTON'S THIRD LAW OF MOTION

20. What is Newton's third law of motion?

____

____

____

21. Explain why Newton's third law can be stated as "all forces act in pairs."

____

____

22. What action and reaction forces are present when you are sitting on a chair?

____

____

____

23. How do action and reaction forces move a swimmer forward in the water?

____

____

____

24. Since all forces act in pairs, what happens when a force is exerted?

____

25. When a ball falls to Earth, why is it hard to see the effect of the reaction force exerted by the ball on Earth?

____

____

____

Name ______________________ Class ______________ Date ____________

Skills Worksheet

Directed Reading A

Section: Momentum

1. Why does it take a large truck longer to stop than a compact car, even though both are traveling at the same velocity?

____

____

MOMENTUM, MASS, AND VELOCITY

2. The product of the mass and velocity of an object is

its ______________________.

3. Why does a fast-moving car have more momentum than a slow-moving car of the same mass?

____

____

4. What is the equation used to calculate momentum?

____

____

Match the correct description with the correct term. Write the letter in the space provided.

______ **5.** the mass of an object in kilograms

______ **6.** the velocity of an object in meters per second

______ **7.** units of momentum

______ **8.** kilograms multiplied by meters per second

a. *m*

b. *v*

c. kg•m/s

d. *p*

9. What is the direction of momentum?

____

____

THE LAW OF CONSERVATION OF MOMENTUM

______ **10.** If a cue ball hits a billiard ball so that the billiard ball starts moving and the cue ball stops, what happens to the cue ball's momentum?
a. Some of the cue ball's momentum has transferred from the billiard ball.
b. All of the cue ball's momentum has transferred from the billiard ball.
c. All of the cue ball's momentum has transferred to the billiard ball.
d. Some of the cue ball's momentum has transferred to the billiard ball.

______ **11.** The law that states that any time objects collide, the total amount of momentum is conserved, or stays the same, is called the
a. law of conservation of momentum.
b. law of preservation of momentum.
c. law of preservation of velocity.
d. law of conservation of velocity.

______ **12.** When two objects stick together, the mass of the combined objects is equal to the
a. mass of the smaller object subtracted from the larger object.
b. product of the masses of the two objects.
c. masses of the two objects added together.
d. mass of the larger object divided by the smaller object.

______ **13.** If momentum is conserved, what happens to velocity when mass changes?
a. Velocity stays the same.
b. Velocity always increases.
c. Velocity always decreases.
d. Velocity changes.

______ **14.** What usually happens to momentum when objects collide?
a. Momentum of each object remains the same.
b. Momentum of each object increases.
c. Momentum of each object becomes equal.
d. Momentum transfers from one object to another.

______ **15.** When objects collide, the total momentum of all objects
a. remains the same. **c.** decreases.
b. increases. **d.** is divided in half.

16. How is the collision of a cue ball and a billiard ball an example of Newton's third law and the conservation of momentum?

____

Name ______________________ Class ______________ Date ____________

Skills Worksheet

Vocabulary and Section Summary

Gravity and Motion

VOCABULARY

In your own words, write a definition of the following terms in the space provided.

1. terminal velocity

____

____

2. free fall

____

____

3. projectile motion

____

____

SECTION SUMMARY

Read the following section summary.

- Gravity causes all objects to accelerate toward Earth at a rate of 9.8 m/s^2.
- Air resistance slows the acceleration of falling objects. An object falls at its terminal velocity when the upward force of air resistance equals the downward force of gravity.
- An object is in free fall if gravity is the only force acting on it.
- Objects in orbit appear to be weightless because they are in free fall.
- A centripetal force is needed to keep objects in circular motion. Gravity acts as a centripetal force to keep objects in orbit.
- Projectile motion is the curved path an object follows when thrown or propelled near the surface of Earth.
- Projectile motion has two components—horizontal motion and vertical motion. Gravity affects only the vertical motion of projectile motion.

Name ______________________ Class ______________ Date ____________

Skills Worksheet

Vocabulary and Section Summary

Newton's Laws of Motion

VOCABULARY

In your own words, write a definition of the following term in the space provided.

1. inertia

____

____

SECTION SUMMARY

Read the following section summary.

- Newton's first law of motion states that the motion of an object will not change if no unbalanced forces act on it.
- Objects at rest will not move unless acted upon by an unbalanced force.
- Objects in motion will continue to move at a constant speed and in a straight line unless acted upon by an unbalanced force.
- Inertia is the tendency of matter to resist a change in motion. Mass is a measure of inertia.
- Newton's second law of motion states that the acceleration of an object depends on its mass and on the force exerted on it.
- Newton's second law is represented by the following equation: $F = m \times a$.
- Newton's third law of motion states that whenever one object exerts a force on a second object, the second object exerts an equal and opposite force on the first object.

Name ______________________ Class ______________ Date ____________

Skills Worksheet

Vocabulary and Section Summary

Momentum

VOCABULARY

In your own words, write a definition of the following term in the space provided.

1. momentum

____

____

SECTION SUMMARY

Read the following section summary.

- Momentum is a property of moving objects.
- Momentum is calculated by multiplying the mass of an object by the object's velocity.
- When two or more objects collide, momentum may be transferred, but the total amount of momentum does not change. This is the law of conservation of momentum.

Name ______________________ Class ______________ Date ______________

Skills Worksheet

Directed Reading A

Section: Fluids and Pressure

1. Any material that can flow and takes the shape of its container is called a(n) ______________________.

2. Name two types of fluids.

____

____

3. What can particles in a fluid do?

____

____

FLUIDS EXERT PRESSURE

4. What happens when you pump up a bicycle tire?

____

____

____

5. The amount of force exerted per unit area of a surface is called ______________________.

6. Force divided by area equals ______________________.

7. The SI unit of pressure is the ______________________.

8. The force of one newton exerted over an area of one square meter is one ______________________.

9. Why does a soap bubble get rounder instead of longer as you blow into it?

____

____

____

ATMOSPHERIC PRESSURE

Match the correct description with the correct term. Write the letter in the space provided.

______ **10.** pressure caused by the weight of the atmosphere

______ **11.** percentage of gases found within 10 km of Earth's surface

______ **12.** force that holds the atmosphere in place

______ **13.** number of newtons pressing on every square centimeter of your body

a. 10
b. atmospheric pressure
c. 80
d. gravity

14. As you travel "deeper" into the atmosphere, how is atmospheric pressure affected?

____

____

____

Number each location listed from 1 to 5 in order of lowest to highest pressure.

______ **15.** Mount Everest's peak

______ **16.** La Paz, Bolivia

______ **17.** airplane at 12,000 m

______ **18.** beach at sea level

______ **19.** space shuttle at 150,000 m above sea level

20. Why do your ears "pop" when you take off in an airplane?

____

____

WATER PRESSURE

______ **21.** Water pressure increases as

a. gravity decreases.
b. air pressure decreases.
c. depth increases.
d. particles collide.

______**22.** Water pressure and atmospheric pressure affect total pressure on objects that are
a. underground.
b. above sea level.
c. in a car.
d. underwater.

______**23.** Water pressure does NOT depend on
a. atmospheric pressure.
b. the amount of fluid present.
c. air pockets.
d. gravity.

______**24.** Water is about 1,000 times more dense than
a. air.
b. pressure.
c. gravity.
d. oil.

25. The amount of matter in a given volume, or mass per unit volume, is called ______________________.

26. Why does water exert more pressure than air?

____

27. The pressure at 500 m below the surface is about ______________________ KPa.

28. The pressure at 8,000 m is about ______________________ KPa.

PRESSURE DIFFERENCES AND FLUID FLOW

29. How does pressure change as you a drink through a straw?

____

____

____

30. What happens when pressure is lower inside the lungs than outside the lungs?

____

31. How do pressure differences affect the direction in which fluids flow?

____

____

32. How do pressure differences explain destructive effects of a tornado's winds?

____

____

Name ______________________ Class ______________ Date ______________

Skills Worksheet

Directed Reading A

Section: Buoyant Force

1. The upward force that fluids exert on all matter is called ______________________.

BUOYANT FORCE AND FLUID PRESSURE

2. In a fluid, buoyant force exists because the pressure is greater at the ______________________ of an object than the pressure at the top.

3. State Archimedes' principle.

____

____

____

____

4. The weight of displaced fluid determines the ______________________ on an object.

WEIGHT VERSUS BUOYANT FORCE

5. If the weight of the water an object displaces is equal to the weight of the object, the object ______________________.

6. If the weight of the water an object displaces is less than the weight of the object, the object ______________________.

7. If the weight of the water an object displaces is greater than the object's weight, the object is ______________________.

Match the correct description with the correct formula. Write the letter in the space provided.

______ **8.** when a rock sinks

______ **9.** when a duck floats

______ **10.** when a fish is suspended in the water

a. Buoyant force is less than weight.

b. Buoyant force equals weight.

c. Buoyant force is greater than weight.

FLOATING, SINKING, AND DENSITY

11. How does the density of a rock affect its ability to float?

____

____

____

____

12. Why does an ice cube float in water?

____

____

____

____

13. Why does a helium balloon float in air?

____

____

____

____

CHANGING OVERALL DENSITY

14. How does the shape of a steel ship allow the ship to float?

____

____

____

____

15. What would happen if a steel ship were NOT hollow?

____

____

____

____

16. Why are ships built to displace more water than necessary for them to float?

____

____

____

____

17. What is the purpose of a submarine's ballast tanks?

____

____

____

____

18. How is compressed air used in a submarine?

____

____

____

____

19. How does a fish's swim bladder affect its overall density?

____

____

____

____

20. How do fish without swim bladders adjust to differences in fluid density?

____

____

____

____

Name ______________________ Class ______________ Date ______________

Skills Worksheet

Directed Reading A

Section: Fluids and Motion

1. What happens if you blow between two sheets of paper held with the flat faces parallel to each other?

____

____

FLUID SPEED AND PRESSURE

______ **2.** What does Bernoulli's principle say about the speed of a moving fluid?

- **a.** The faster the fluid's speed is, the higher the pressure.
- **b.** The slower the fluid's speed is, the lower the pressure.
- **c.** The faster the fluid's speed is, the lower the pressure.
- **d.** Speed and pressure are not related.

3. When you blow between two sheets of paper held parallel to each other, how is Bernoulli's principle at work?

____

____

4. Why would a table-tennis ball attached to a string stay in the water stream under a faucet?

____

____

FACTORS THAT AFFECT FLIGHT

5. The fast-moving air above an airplane wing exerts less pressure than the slow-moving air below the wing, according to ______________________.

6. The upward force acting on an airplane wing due to air flow is called ______________________.

7. The forward force produced by a plane's engine is called ______________________.

8. How does thrust increase lift?

____

____

9. A jet plane needs ______________________ wings.

10. A glider needs ______________________ wings.

11. How does wing size affect how birds fly?

____

____

12. When a screwball is thrown, air speed and pressure react according to ______________________.

DRAG AND MOTION IN FLUIDS

13. The force that opposes or restricts motion in a fluid is called ______________________.

14. An irregular or unpredictable flow of fluids is known as ______________________.

15. How does drag affect an aircraft's speed, and how can drag be reduced?

____

____

16. How do birds respond to turbulence?

____

____

PASCAL'S PRINCIPLE

17. Define Pascal's principle.

____

____

____

18. Pascal's principle is used by ______________________ devices to move or lift objects.

19. How do hydraulic devices make use of fluid pressure?

____

____

20. Hydraulic devices can ______________________ forces.

21. When brakes are used to stop a car, ______________________ is in effect.

Name ______________________ Class ______________ Date ____________

Skills Worksheet

Vocabulary and Section Summary

Fluids and Pressure

VOCABULARY

In your own words, write a definition of the following terms in the space provided.

1. fluid

____

____

2. pressure

____

____

3. pascal

____

____

4. atmospheric pressure

____

____

SECTION SUMMARY

Read the following section summary.

- A fluid is any material that flows and takes the shape of its container.
- Pressure is force exerted on a given area.
- Moving particles of matter create pressure by colliding with one another and with the walls of their container.
- The pressure caused by the weight of the atmosphere is called *atmospheric pressure.*
- Fluid pressure increases as depth increases.
- As depth increases, water pressure increases faster than atmospheric pressure does because water is denser than air.
- Fluids flow from areas of high pressure to areas of low pressure.

Name ______________________ Class ______________ Date ______________

Skills Worksheet

Vocabulary and Section Summary

Buoyant Force

VOCABULARY

In your own words, write a definition of the following terms in the space provided.

1. buoyant force

____

____

2. Archimedes' principle

____

____

SECTION SUMMARY

Read the following section summary.

- All fluids exert an upward force called *buoyant force.*
- Buoyant force is caused by differences in fluid pressure.
- Archimedes' principle states that the buoyant force on an object is equal to the weight of the fluid displaced by the object.
- Any object that is more dense than the surrounding fluid will sink. An object that is less dense than the surrounding fluid will float.
- The overall density of an object can be changed by changing the object's shape, mass, or volume.

Name ______________________ Class ______________ Date ______________

Skills Worksheet

Vocabulary and Section Summary

Fluids and Motion

VOCABULARY

In your own words, write a definition of the following terms in the space provided.

1. Bernoulli's principle

____

____

2. lift

____

____

3. thrust

____

____

4. drag

____

____

5. Pascal's principle

____

____

SECTION SUMMARY

Read the following section summary.

- Bernoulli's principle states that fluid pressure decreases as the speed of the fluid increases.
- Wing shape allows airplanes to take advantage of Bernoulli's principle to achieve flight.
- Lift on an airplane is determined by wing size and thrust.
- Drag opposes motion through fluids.
- Pascal's principle states that a change in pressure in an enclosed fluid is transmitted equally to all parts of the fluid.

Name ______________________ Class ______________ Date ____________

Skills Worksheet

Directed Reading A

Section: Work and Power

______ **1.** What is the transfer of energy to an object using a force that causes the object to move in the direction of the force?

a. movement
b. power
c. work
d. force

WHAT IS WORK?

______ **2.** Which of the following is considered work?

a. throwing a bowling ball
b. doing homework
c. watching television
d. trying to push a box, but not moving it

3. One way you can tell that the bowler has done work is that when the ball is moving, it has ______________________ energy.

4. When a bowling ball has kinetic energy, the bowler has transferred ______________________ to the ball.

5. What two things need to happen for work to be done on an object?

____

____

____

____

HOW MUCH WORK?

6. Why is it the same amount of work for a hiker to climb straight up a cliff and to walk up a slope?

____

____

____

____

7. The formula used to calculate work is:

work = ______________________ × ______________________.

8. The unit used to express energy is the ______________________.

9. Work is the transfer of ______________________ to an object.

10. Increasing the amount of work done can be accomplished by increasing what two things?

____

____

POWER: HOW FAST WORK IS DONE

_____ **11.** What is the rate at which work is done or energy is transformed called?

a. force
b. power
c. work
d. energy

_____ **12.** What is the equation used to calculate power?

a. $t = \frac{P}{W}$
b. $P = \frac{W}{t}$
c. $t = \frac{W}{P}$
d. $W = \frac{t}{P}$

_____ **13.** What is the unit used to express power called?

a. joule
b. inch
c. watt
d. meter

_____ **14.** One watt is equal to

a. one joule per hour.
b. one joule per minute.
c. one joule per day.
d. one joule per second.

15. Name the two things that power measures.

____

____

16. In what two instances does power output become greater?

____

____

17. If you sand a shelf by hand, the energy needed is the same as if you sanded it with an electric sander, but the power output is ____________________.

18. How does a powerful engine affect the performance of a car?

____

____

____

____

Name ______________________ Class ______________ Date ____________

Skills Worksheet

Directed Reading A

Section: What Is a Machine?

MACHINES: MAKING WORK EASIER

______ **1.** What is a device that makes work easier by changing the size or direction of force?

a. a machine
b. a load
c. an engine
d. a computer

2. Name three examples of everyday machines.

____

______ **3.** What type of common machine is a screwdriver that is used to pry off the lid on a paint can?

a. a pulley
b. a wheel
c. a lever
d. a screw

4. The work you do on a machine is called ______________________.

5. The work done by a machine on an object is called

______________________.

6. Work output can never be greater than ______________________.

7. Why do machines need less force to do the same amount of work?

____

8. When a screwdriver is used to open a can, both the size and direction of the ______input______ change.

9. A ramp will decrease the size of the input force needed to lift a box and ______________________ the distance over which the force is exerted.

10. When a machine changes the size of the force, the ______________________ through which the force is exerted must also change.

MECHANICAL ADVANTAGE

______ **11.** What is the number of times a machine multiplies force called?

a. output force
b. input force
c. mechanical advantage
d. work output

______ **12.** Which of the following is the formula for finding mechanical advantage?

a. *MA = input force ÷ output force*
b. *MA = output force ÷ input force*
c. *MA = input force ÷ output force × 100*
d. *MA = output force ÷ input force × 100*

13. A machine that has a mechanical advantage of greater than 1 has an output force that is ____________________ than the input force.

14. A machine that has a mechanical advantage of less than 1 reduces the output force but can increase the distance an object moves.

MECHANICAL EFFICIENCY

______ **15.** What is the quantity that measures the ratio of work output to work input called?

a. mechanical work
b. mechanical efficiency
c. mechanical force
d. mechanical energy

______ **16.** Which of the following is the equation for finding mechanical efficiency?

a. *mechanical efficiency = work input ÷ work output*
b. *mechanical efficiency = work output ÷ work input*
c. *mechanical efficiency = work input ÷ work output × 100*
d. *mechanical efficiency = work output ÷ work input × 100*

17. When a machine drills holes in metal, some of the work input is used to overcome ____________________ between the metal and the drill.

18. What would a machine that had 100% mechanical efficiency be called?

____

19. Why is it impossible to build an ideal machine?

____

20. What do some machines use to lower friction between moving parts?

____

Name ______________________ Class ______________ Date ______________

Skills Worksheet

Directed Reading A

Section: Types of Machines

1. A knife is actually a very sharp ______________________.

2. What are the six simple machines that all other machines are made from?

____

____

____

LEVERS

______ 3. A simple machine with a bar that pivots at a fixed point is a(n)
- a. wedge.
- b. lever.
- c. knife.
- d. screw.

______ 4. What is the fixed point on a lever called?
- a. bolt
- b. pivot point
- c. fulcrum
- d. wedge

______ 5. What do first-class levers always change the direction of?
- a. input force
- b. output force
- c. distance
- d. fulcrum

______ 6. When you use the claw end of a hammer to remove a nail, what type of simple machine are you using?
- a. wedge
- b. first-class lever
- c. screw
- d. pulley

7. The three classes of lever are based on the location of what three features?

____

____

____

8. Where are the fulcrum, the load, and the input force located in a first-class lever, a second-class lever, and a third-class lever?

____

____

____

9. In a second-class lever, why must you exert input force over a greater distance?

____

____

10. Why is the output force always less than the input force in a third-class lever?

____

____

PULLEYS

______ **11.** Which of the following simple machines has a grooved wheel that holds a rope or cable?

a. lever
b. wedge
c. pulley
d. wheel and axle

______ **12.** Which type of pulley is attached to something that does not move?

a. fixed pulley
b. movable pulley
c. block and tackle
d. simple pulley

______ **13.** Which type of pulley is attached to the object being moved?

a. fixed pulley
b. movable pulley
c. block and tackle
d. simple pulley

______ **14.** What determines the mechanical advantage of a block and tackle?

a. the amount of input force
b. the amount of output force
c. the weight of the rope
d. the number of rope segments

15. How does a fixed pulley affect force?

____

____

16. How does a movable pulley move?

____

____

17. Describe a block and tackle.

____

____

draw the pic. too

WHEEL AND AXLE

_____ **18.** A faucet is what type of simple machine?

a. lever
b. pulley
c. wheel and axle
d. wedge

19. What does a wheel and axle consist of?

____

____

20. How do you find the mechanical advantage of a wheel and axle?

____

____

INCLINED PLANES

_____ **21.** inclined plane

_____ **22.** screw

_____ **23.** wedge

a. a simple machine that consists of an inclined plane wrapped around a cylinder

b. a simple machine that is made up of two inclined planes and that moves

c. a simple machine that is a straight, slanted surface, which facilitates raising a load

24. How do you calculate the mechanical advantage of an inclined plane?

____

25. Name three examples of a wedge.

____

____

____

26. How do you find the mechanical advantage of a wedge?

____

27. What happens when a screw is turned?

____

28. The longer the spiral on a screw is and the closer together the threads are, the greater the screw's ____________________.

COMPOUND MACHINES

29. A machine that is made of more than one simple machine is

a(n) ________________________.

30. What three simple machines make up a can opener?

____

____

____

31. Why is the mechanical efficiency of most compound machines lower than most simple machines?

____

____

32. Name two compound machines.

____

____

33. Why is it important to reduce friction on compound machines?

____

____

____

Name ______________________ Class ______________ Date ____________

Skills Worksheet

Vocabulary and Section Summary

Work and Power

VOCABULARY

In your own words, write a definition of the following terms in the space provided.

1. work

____

____

2. joule

____

____

3. power

____

____

4. watt

____

____

SECTION SUMMARY

Read the following section summary.

- In scientific terms, *work* is done when a force causes an object to move in the direction of the force.
- Work is calculated as force times distance. The unit of work is the newton-meter, or joule.
- *Power* is a measure of how fast work is done.
- Power is calculated as work divided by time. The unit of power is the joule per second, or watt.

Name ______________________ Class ______________ Date ____________

Skills Worksheet

Vocabulary and Section Summary

What Is a Machine?

VOCABULARY

In your own words, write a definition of the following terms in the space provided.

1. machine

____

____

2. work input

____

____

3. work output

____

____

4. mechanical advantage

____

____

5. mechanical efficiency

____

____

SECTION SUMMARY

Read the following section summary.

- A machine makes work easier by changing the size or direction (or both) of a force.
- A machine can increase force or distance, but not both.
- Mechanical advantage tells how many times a machine multiplies force.
- Mechanical efficiency is a comparison of a machine's work output with work input.
- Machines are not 100% efficient because some of the work done is used to overcome friction.

Name ______________________ Class ______________ Date ____________

Skills Worksheet

Vocabulary and Section Summary

Types of Machines

VOCABULARY

In your own words, write a definition of the following terms in the space provided.

1. lever

____

____

2. pulley

____

____

3. wheel and axle

____

____

4. inclined plane

____

____

5. wedge

____

____

6. screw

____

____

7. compound machine

____

____

SECTION SUMMARY

Read the following section summary.

- In a first-class lever, the fulcrum is between the force and the load. In a second-class lever, the load is between the force and the fulcrum. In a third-class lever, the force is between the fulcrum and the load.
- The mechanical advantage of an inclined plane is length divided by height. Wedges and screws are types of inclined planes.
- A wedge is a type of inclined plane. Its mechanical advantage is its length divided by its greatest thickness.
- The mechanical advantage of a wheel and axle is the radius of the wheel divided by the radius of the axle.
- Types of pulleys include fixed pulleys, movable pulleys, and block and tackles.
- Compound machines consist of two or more simple machines.
- Compound machines have low mechanical efficiencies because they have more moving parts and therefore more friction to overcome.

Name ______________________ Class ______________ Date ____________

Skills Worksheet

Directed Reading A

Section: What Is Energy?

ENERGY AND WORK: WORKING TOGETHER

______ **1.** What is the ability to do work called?
- **a.** movement
- **b.** energy
- **c.** power
- **d.** force

2. Work is a transfer of ______________________.

3. How is energy transferred when one object does work on another?

____

4. What units are used to express this energy transfer?

____

KINETIC ENERGY

______ **5.** Which of the following is the energy of motion?
- **a.** potential energy
- **b.** mechanical energy
- **c.** kinetic energy
- **d.** gravitational energy

______ **6.** In the formula for kinetic energy, what does the m stand for?
- **a.** more
- **b.** moving
- **c.** mass
- **d.** meter

7. How does increasing mass affect kinetic energy?

____

8. Why are car crashes more dangerous at higher speeds than at lower speeds?

____

____

POTENTIAL ENERGY

9. The energy an object has because of its position is called

________________________ energy.

10. When you lift an object, energy is transferred to the object, which gives the object ________________________.

11. The amount of gravitational potential energy that an object has depends on its weight and ________________________.

12. What formula is used to calculate gravitational potential energy?

____

13. The amount of force that must be used on an object to lift it is

________________________.

14. What is an object's height a measure of?

____

MECHANICAL ENERGY

______ **15.** Which of the following types of energy equals the total energy of motion and position?

a. mechanical energy
b. kinetic energy
c. potential energy
d. moving energy

16. What is the formula used to find mechanical energy?

____

17. The juggler moves the pin with his hand and gives ________________________ energy to the pin.

18. As the juggler's pin leaves his hand, the pin's kinetic energy begins to change to ________________________ energy.

19. How can you tell that the kinetic energy is decreasing as the juggler's pin rises?

____

OTHER FORMS OF ENERGY

Match the correct description with the correct term. Write the letter in the space provided. Some terms will not be used.

__a__ **20.** energy caused by an object's vibrations

__c__ **21.** energy that comes from changes in the nucleus of an atom

__g__ **22.** all of the kinetic energy due to random motion of the particles that make up an object

__e__ **23.** energy of moving electrons

__b__ **24.** energy of a compound that changes as its atoms are rearranged

__d__ **25.** energy produced by the vibrations of electrically charged particles

a. sound energy
b. chemical energy
c. nuclear energy
d. light energy
e. electrical energy
f. mechanical energy
g. thermal energy

26. How do particles move at higher temperatures compared to how they move at lower temperatures?

____

27. Chemical energy is a form of ______________________ energy because it depends on the position and arrangement of the atoms in a compound.

28. How is electrical energy produced at power plants?

____

____

29. When you stretch a guitar string, what kind of energy does the string store?

____

30. When you release a guitar string, what kind of energy makes the string vibrate?

____

31. What form of energy can travel through a vacuum?

____

32. What is the difference between fission and fusion?

____

____

Name ______________________ Class ______________ Date ____________

Skills Worksheet

Directed Reading A

Section: Energy Conversions

1. A change from one form of energy to another is called

a(n) ________________________

KINETIC ENERGY AND POTENTIAL ENERGY

______ **2.** When the skateboarder reaches the top of the half-pipe, which of the following types of energy is at its maximum?
a. mechanical energy
b. kinetic energy
c. potential energy
d. elastic potential energy

______ **3.** As the skateboarder speeds down through the bottom of the half-pipe, which of the following types of energy is at its maximum?
a. mechanical energy
b. kinetic energy
c. potential energy
d. elastic potential energy

______ **4.** Which of the following types of energy is present in the wound-up rubber band in a toy airplane?
a. mechanical energy
b. kinetic energy
c. potential energy
d. elastic potential energy

5. When the rubber band on the airplane is released, the stored energy

becomes ________________________ energy, spinning the propeller.

6. A stretched rubber band stores ________________________ energy.

CONVERSIONS INVOLVING CHEMICAL ENERGY

______ **7.** Which of the following types of energy comes from the food you eat?
a. chemical energy
b. thermal energy
c. light energy
d. nuclear energy

______ **8.** When you are active, chemical energy of food is converted into which of the following types of energy?
a. kinetic energy
b. thermal energy
c. mechanical energy
d. potential energy

______ **9.** Which of the following types of energy from food is used to maintain body temperature?
a. chemical energy
b. thermal energy
c. light energy
d. nuclear energy

______ **10.** Which of the following types of energy do plants use to make chemical energy?
a. gravitational energy
b. thermal energy
c. light energy
d. nuclear energy

11. When you eat food, you are taking in energy that first came from the ______________________.

12. Define photosynthesis.

____

____

____

13. How is the chemical energy from a tree converted into thermal energy?

____

____

____

____

WHY ENERGY CONVERSIONS ARE IMPORTANT

14. Give two examples of energy conversions that take place within a hair dryer.

____

____

____

15. Name two common energy conversions that involve electrical energy.

____

____

____

____

ENERGY AND MACHINES

16. Work is made easier when a(n) ______________________ changes the direction or size of the force needed to do the work.

17. When riding a bike, your legs transfer ______________________ energy to the pedals by pushing them around in a circle.

18. Energy from the sun is measured using a(n)______________________.

19. In a radiometer, which vanes absorb the most light energy? What happens after the light is absorbed?

____

____

____

____

Name ______________________ Class ______________ Date ____________

Skills Worksheet

Directed Reading A

Section: Conservation of Energy

WHERE DOES THE ENERGY GO?

_____ **1.** Which of the following forces opposes motion between two surfaces that are touching?
a. friction
b. kinetic energy
c. potential energy
d. current

_____ **2.** When a roller coaster moves, which of the following helps to overcome friction?
a. friction
b. current
c. energy
d. movement

_____ **3.** Which of the following causes some of the potential energy of a moving roller coaster to be converted to thermal energy?
a. movement
b. thermal energy
c. friction
d. sound energy

4. Name two locations where there is friction on a roller coaster.

____

____

5. On a roller coaster, where is the potential energy the greatest?

____

6. The kinetic energy at the bottom of the first hill on a roller coaster is less than the ______________________ energy at the top of the hill.

ENERGY IS CONSERVED WITHIN A CLOSED SYSTEM

7. When a group of objects transfer energy only to each other, a(n) ______________________ is created.

8. On a roller coaster, some mechanical energy is always converted into ______________________ energy.

9. What three types of energy from a roller coaster are added together to equal the total amount of original potential energy?

____

____

____

10. The rule that energy cannot be created or destroyed is explained by the _______________.

11. The total amount of _______________ in a closed system is always the same regardless of how many conversions take place.

12. In a light bulb, some energy is converted to _______________ energy, making the bulb feel warm.

NO CONVERSION WITHOUT THERMAL ENERGY

13. Any time energy is converted into another form of energy, some of the original energy is converted into _______________ energy.

14. The thermal energy due to _______________ is not useful energy.

15. What happens to the wasted thermal energy in a car?

____

____

16. A machine that puts out exactly as much energy as it takes in is a(n) _______________.

17. What is needed to keep a machine moving?

____

18. What energy does the "drinking bird" use to evaporate water from its head?

____

19. Define energy efficiency.

____

____

20. New cars have a smooth _______________ that reduces friction between the body of the car and the air.

21. A car with high _______________ can go farther than other cars with the same amount of gas.

22. There is less wasted energy when energy _______________ are more efficient.

Name ______________________ Class ______________ Date ______________

Skills Worksheet

Directed Reading A

Section: Energy Resources

1. A natural resource that can be converted into other forms of energy to do work is called a(n) ______________________.

2. The source responsible for most other energy resources is the ______________________.

NONRENEWABLE RESOURCES

______ **3.** Which of the following energy resources cannot be replaced or is replaced more slowly than it is used?
a. usable resource
b. reusable resource
c. nonrenewable resource
d. energy resource

______ **4.** Which of the following energy resources are formed from buried plants and animals?
a. fossil fuels
b. fossil resources
c. renewable fuels
d. animal fuels

5. What are the three most common fossil fuels?

____

____

6. The plants and animals that form fossil fuels stored ______________________ from the sun.

7. A common use of coal in the United States is to generate ______________________.

8. Name three products that come from petroleum.

____

____

9. Name two uses for natural gas.

____

____

10. To convert the chemical energy in fossil fuels into electric energy, a(n) ______________________ is used.

11. In an electric generator, a magnet spins to generate

a(n) ______________________ in the wire.

12. A nuclear power plant generates ______________________ energy that boils water to produce steam.

13. The spinning generator of a nuclear power plant converts

______________________ energy into electrical energy.

14. Nuclear energy is generated from ______________________ elements.

15. The nucleus of a uranium atom is split into two to release nuclear energy in a

process called ______________________.

16. Why is nuclear energy considered a nonrenewable resource?

____

____

RENEWABLE RESOURCES

______ **17.** Of the following energy resources can be replaced over a relatively short period of time?

a. usable resource
b. reusable resource
c. nonrenewable resource
d. renewable resource

18. Name three types of renewable resources.

____

____

____

19. Solar energy can be changed into electrical energy through

______________________.

20. Through what can houses passively collect solar energy?

____

21. The potential energy of water in a reservoir can be changed into

______________________ energy as the water flows downhill.

22. How does a hydroelectric dam change kinetic energy into electrical energy?

____

____

23. What causes wind?

____

24. What kind of energy of wind turns the blades of a windmill?

____

25. Wind turbines convert the kinetic energy of air into ________________________ energy by turning a generator.

26. The thermal energy caused by the heating of Earth's crust is called ________________________.

27. Some geothermal power plants pump water underground next to hot ________________________.

28. Organic matter that can be burned to release energy is called ________________________.

THE TWO SIDES TO ENERGY RESOURCES

29. Name one disadvantage of fossil fuels.

____

30. Why can't solar energy be used to meet the energy needs of large cities?

____

____

31. Why is hydroelectric energy not always a desirable energy resource?

____

____

Name ______________________ Class ______________ Date ______________

Skills Worksheet

Vocabulary and Section Summary

What Is Energy?

VOCABULARY

In your own words, write a definition of the following terms in the space provided.

1. energy

____

____

2. kinetic energy

____

____

3. potential energy

____

____

4. mechanical energy

____

____

SECTION SUMMARY

Read the following section summary.

- Energy is the ability to do work, and work equals the transfer of energy. Energy and work are expressed in units of joules (J).
- Kinetic energy is energy of motion and depends on speed and mass.
- Potential energy is energy of position. Gravitational potential energy depends on weight and height.
- Mechanical energy is the sum of kinetic energy and potential energy.
- Thermal energy and sound energy can be considered forms of kinetic energy.
- Chemical energy, electrical energy, and nuclear energy can be considered forms of potential energy.

Name ______________________________ Class ______________ Date ____________

Skills Worksheet

Vocabulary and Section Summary

Energy Conversions

VOCABULARY

In your own words, write a definition of the following term in the space provided.

1. energy conversion

____

____

SECTION SUMMARY

Read the following section summary.

- An energy conversion is a change from one form of energy to another. Any form of energy can be converted into any other form of energy.
- Kinetic energy is converted to potential energy when an object is moved against gravity.
- Elastic potential energy is another example of potential energy.
- Your body uses the food you eat to convert chemical energy into kinetic energy.
- Plants convert light energy into chemical energy.
- Machines can transfer energy and can convert energy into a more useful form.

Name ______________________ Class ______________ Date ____________

Skills Worksheet

Vocabulary and Section Summary

Conservation of Energy

VOCABULARY

In your own words, write a definition of the following terms in the space provided.

1. friction

____

____

2. law of conservation of energy

____

____

SECTION SUMMARY

Read the following section summary.

- Because of friction, some energy is always converted into thermal energy during an energy conversion.
- Energy is conserved within a closed system. According to the law of conservation of energy, energy cannot be created or destroyed.
- Perpetual motion is impossible because some of the energy put into a machine is converted into thermal energy because of friction.

Name ______________________ Class ______________ Date ____________

Skills Worksheet

Vocabulary and Section Summary

Energy Resources

VOCABULARY

In your own words, write a definition of the following terms in the space provided.

1. nonrenewable resource

____

____

2. fossil fuel

____

____

3. renewable resource

____

____

SECTION SUMMARY

Read the following section summary.

- An energy resource is a natural resource that can be converted into other forms of energy in order to do useful work.
- Nonrenewable resources cannot be replaced after they are used or can be replaced only after long periods of time. They include fossil fuels and nuclear energy.
- Renewable resources can be replaced in nature over a relatively short period of time. They include energy from the sun, wind, and water; geothermal energy; and biomass.
- The sun is the source of most energy on Earth.
- Choices about energy resources depend on where you live and what you need energy for.

Name ______________________ Class ______________ Date ____________

Skills Worksheet

Directed Reading A

Section: Temperature

1. Why must you use temperature to specify how hot or cold something is?

____

____

WHAT IS TEMPERATURE?

______ 2. Temperature is a measure of which property of an object's particles?
- **a.** average potential energy
- **b.** average mechanical energy
- **c.** average kinetic energy
- **d.** average volume

3. As particles in an object move faster, they have more

_____________________, so that the object's temperature is higher.

4. Particles of matter move at different speeds, so when you measure an object's

temperature, you measure the _____________________ of its particles.

5. How does the amount of a substance affect its temperature? Explain your answer.

____

____

MEASURING TEMPERATURE

______ 6. To measure the temperature of a cup of hot chocolate, you would
- **a.** touch it with your finger.
- **b.** put a thermometer in it.
- **c.** take a sip of it.
- **d.** look at the steam rising from it.

______ 7. Mercury and alcohol are used in thermometers because
- **a.** they remain liquid over a large temperature range.
- **b.** they freeze and boil at the same temperatures as water does.
- **c.** they are cheaper to use than other substances.
- **d.** they are safer to use than other substances.

8. The increase in volume of a substance due to an increase in the temperature

of the object is called_____________________.

9. Explain how temperature units (parts) of the Celsius scale are determined. What are the units called?

____

____

10. The temperatures 212°F, 100°C, and 373 K are ______________________.

11. The official SI temperature scale uses units called ______________________.

12. The temperature at which all molecular motion stops is

called ______________________.

13. What equation would you use to convert temperature in degrees Fahrenheit (°F) to degrees Celsius (°C)?

____

14. Describe how a change in temperature of 1° differs on the Fahrenheit, Celsius, and Kelvin temperature scales.

____

____

MORE ABOUT THERMAL EXPANSION

Match the correct description with the correct term. Write the letter in the space provided.

__c__ **15.** a thin strip of two different metals that coils and uncoils in response to temperature changes

__d__ **16.** a thin glass tube filled with a liquid that measures temperature

__b__ **17.** a device that controls the heater in a home

__a__ **18.** a gap in the pavement of a bridge that allows the bridge to expand without breaking

a. expansion joint
b. thermostat
c. bimetallic strip
d. thermometer

19. If the weather is very hot, the pavement of a bridge can heat up enough so that ______________________ takes place.

20. Why are two different metals needed when making a bimetallic strip for a home thermostat?

____

____

____

21. How does thermal expansion make a hot-air balloon rise?

____

____

____

Name ______________________________ Class ______________ Date ____________

Skills Worksheet

Directed Reading A

Section: What Is Heat?

1. Why does a stethoscope feel cold?

____

TRANSFERRED THERMAL ENERGY

______ **2.** Under what condition can heat pass between two objects?
- **a.** The objects must both be hot.
- **b.** The objects must both be large.
- **c.** The objects must be at different temperatures.
- **d.** The objects must have a lot of energy.

______ **3.** What happens if two objects come in contact with each other and one object is at a higher temperature than the other object?
- **a.** The temperatures of both objects decrease.
- **b.** The temperatures of both objects increase.
- **c.** Energy is transferred from the object with a lower temperature.
- **d.** Energy is transferred from the object with a higher temperature.

______ **4.** If a large pan of soup and a small bowl of soup have the same temperature, what do you know about the thermal energy of the two containers of soup?
- **a.** The large pan of soup has more thermal energy.
- **b.** The small bowl of soup has more thermal energy.
- **c.** The pan and bowl of soup have the same thermal energy.
- **d.** Not enough is known about either object's thermal energy to say.

5. When two objects are touching each other and are at the same temperature, there is no net change in either object's ______________________.

CONDUCTION, CONVECTION, AND RADIATION

______ **6.** Which of the following is NOT a form of thermal energy transfer?
- **a.** conduction
- **b.** conversion
- **c.** convection
- **d.** radiation

______ **7.** Thermal conduction occurs when particles with higher average kinetic energies transfer energy
a. through collision to particles with higher kinetic energies.
b. through collision to particles with lower kinetic energies.
c. through fluid movement to particles with higher kinetic energies.
d. through fluid movement to particles with lower kinetic energies.

______ **8.** Substances that do not conduct thermal energy very well are called
a. thermal convectors.
b. thermal conductors.
c. thermal exchangers.
d. thermal insulators.

9. What are materials that transfer thermal energy well called?

____

10. Transfer of thermal energy by the movement of a liquid or a gas is called ______________________.

11. When you boil water, the water near the stove burner becomes less ______________________ because the temperature of the particles increases and, as a result of their increased energy, they spread apart.

12. The circular motion of a liquid or gas due to a density difference that results from temperature differences within the liquid or gas is called a(n) ______________________.

13. The transfer of thermal energy by electromagnetic waves is called ______________________.

14. Energy transfer across empty space involves ______________________, such as visible light and infrared waves.

15. Your body feels warmer when it absorbs ______________________ waves.

16. Explain how certain greenhouse gases cause the temperature of Earth's atmosphere to increase.

____

____

____

____

17. Explain how the greenhouse effect in Earth's atmosphere can be both helpful and harmful.

____

____

____

____

HEAT AND TEMPERATURE CHANGE

18. The rate at which a substance conducts thermal energy is

called _______________.

19. Explain why a metal seat belt buckle is hotter to the touch than the cloth of the seat belt when both have been exposed to the same amount of sunlight for a long time.

____

____

____

____

20. When equal amounts of energy are transferred to or from equal masses of different substances, the substances will undergo different changes in

temperature because of their different _______________.

21. Specific heat is the amount of energy needed to change the temperature

of _______________ (amount) of a substance

by _______________.

22. Explain why water does not heat up or cool off as quickly as air does.

____

____

____

____

23. The specific heat values of most ______________________ tend to be quite low.

24. To calculate the amount of energy transferred by heat to or from an object, you must know the mass of the object, the change in the object's ______________________, and its ______________________.

25. When the value for heat is ______________________, it means that energy has been transferred from an object, so that the object's temperature has decreased.

Name ______________________ Class ______________ Date ____________

Skills Worksheet

Directed Reading A

Section: Matter and Heat

1. Why does a frozen juice bar melt on a hot day before you have finished eating it?

____

____

____

____

STATES OF MATTER

2. The physical forms in which a substance can exist are called

______________________.

______ 3. The state of matter of a substance does not depend on
  a. the speeds of the particles in the substance.
  b. the masses of the particles in the substance.
  c. the attraction between the particles in the substance.
  d. the pressure around the particles in the substance.

______ 4. If you have equal masses of a substance in each of its three familiar states and each at a different temperature, the substance will have the least thermal energy as a
  a. plasma.
  b. gas.
  c. liquid.
  d. solid.

5. Why does a substance have more thermal energy as a gas than as a liquid or as a solid?

____

____

Match the correct description with the correct term. Write the letter in the space provided.

______ 6. The particles of a substance are able to slide past one another.

______ 7. The particles of a substance move independently of one another.

______ 8. The particles of a substance vibrate in place.

a. solid
b. liquid
c. gas

CHANGES OF STATE

______ **9.** What is a change in a substance from one state of matter to another called?
- **a.** a chemical change
- **b.** a change of state
- **c.** a physical property
- **d.** a change of identity

______ **10.** A change of state changes
- **a.** the physical properties of a substance.
- **b.** the chemical properties of a substance.
- **c.** the melting point of a substance.
- **d.** the boiling point of a substance.

______ **11.** When a gas changes to a liquid, the change of state is called
- **a.** freezing
- **b.** melting.
- **c.** boiling.
- **d.** condensing.

______ **12.** When a solid changes to a liquid, the change of state is called
- **a.** freezing.
- **b.** melting.
- **c.** boiling.
- **d.** condensing.

13. Why does the temperature of a substance remain constant when the substance undergoes a change of state?

____

____

____

____

HEAT AND CHEMICAL CHANGES

14. New substances are formed during a(n) ______________________.

15. How does thermal energy cause substances to undergo chemical change? Explain your answer.

____

____

____

16. Is energy always released during a chemical change? Explain your answer.

____

____

____

17. The unit of energy used to measure chemical energy in food is called

a(n) _______________.

18. A Calorie is equivalent to _______________ J.

19. A device used to measure heat is called a(n)_______________.

20. Describe how a bomb calorimeter is used to determine the energy contained in a food sample.

____

____

____

____

Name ______________________ Class ______________ Date ____________

Skills Worksheet

Directed Reading A

Section: Heat Technology

1. Besides a heater, name three other examples of heat technology.

____

____

HEATING SYSTEMS

2. A hot-water heating system makes use of the high

______________________ of water to heat the rooms in a building.

3. In both hot-water and warm-air heating systems, thermal energy spreads

through the air in rooms by means of ______________________.

4. In a warm-air heating system, warm air travels from the furnace

through ______________________ to different rooms.

5. A material that reduces the amount of thermal energy needed to heat or cool

a building is called ______________________.

6. In what ways does a passive solar heating system differ from an active solar heating system?

____

____

____

Match the correct description with the correct term. Write the letter in the space provided.

______ 7. pumps heated water through the pipes of an active solar heating system

______ 8. collects the cooled water in an active solar heating system

______ 9. uses sunlight to heat the water in an active solar heating system

______ 10. blows air over hot-water pipes to transfer thermal energy to the air.

a. solar collector

b. pump

c. fan

d. water storage tank

HEAT ENGINES

_____ **11.** What is a heat engine?
- **a.** an engine that makes fuel
- **b.** a machine that changes heat into mechanical energy
- **c.** a machine used to generate electricity
- **d.** an engine that uses a flywheel

_____ **12.** In the process of combustion,
- **a.** fuel combines with oxygen to produce thermal energy.
- **b.** fuel combines with steam to produce thermal energy.
- **c.** air combines with water to produce steam.
- **d.** air combines with oxygen to produce steam.

_____ **13.** A heat engine that burns fuel inside the engine is called a(n)
- **a.** electric engine.
- **b.** external combustion engine.
- **c.** internal combustion engine.
- **d.** difference engine.

14. One example of an external combustion engine is a(n) ______________________, in which fuel is burned outside a boiler to heat water.

15. In a steam engine, the steam produced expands and pushes a piston,which can be attached to other parts of the machine that do ______________________.

Match the correct description with the correct term. Write the letter in the space provided.

_____ **16.** The piston is pulled down, drawing the mixture of air and gasoline into the cylinder.

_____ **17.** The piston is pushed up, compressing the fuel mixture.

_____ **18.** The compressed fuel mixture is ignited, causing the mixture to burn and the product gases to expand and push the piston down.

_____ **19.** The piston is moved up, pushing the exhaust gases out of the cylinder.

- **a.** compression stroke
- **b.** exhaust stroke
- **c.** intake stroke
- **d.** power stroke

COOLING SYSTEMS

______**20.** An air conditioner is a cooling system that transfers
- **a.** thermal energy from a cool area outside to a warm area inside.
- **b.** specific heat from a cool area outside to a warm area inside.
- **c.** thermal energy from a warm area inside to a warmer area outside.
- **d.** specific heat from a warm area inside to a cool area outside.

______**21.** To transfer thermal energy from an area at a lower temperature to an area at a higher temperature, a cooling system
- **a.** must do work.
- **b.** must have a large thermal conductivity.
- **c.** must have a high specific heat.
- **d.** must produce a large amount of thermal energy.

______**22.** Work in a cooling system is done by the
- **a.** refrigerant.
- **b.** condenser coils.
- **c.** compressor.
- **d.** thermal energy.

23. A gas that has a boiling point below room temperature, which allows it to condense easily, is called a(n)______________________.

24. Explain why the back of a refrigerator feels warm.

____

____

____

HEAT TECHNOLOGY AND THERMAL POLLUTION

______**25.** Thermal pollution is

a. the excessive heating of the atmosphere.
b. the excessive heating of a body of water.
c. the excessive heating of an area of ground.
d. the excessive heating of a large work place.

______**26.** The increased temperature of water that is heated in a power plant and returned to lakes and streams

a. increases the population of plants in the bodies of water.
b. causes flooding.
c. contaminates the water.
d. causes harm to animals living in the bodies of water.

27. What can be done so that power plants reduce thermal pollution?

____

____

____

____

____

____

Name ______________________ Class ______________ Date ____________

Skills Worksheet

Vocabulary and Section Summary

Temperature

VOCABULARY

In your own words, write a definition of the following terms in the space provided.

1. temperature

____

____

2. thermal expansion

____

____

3. absolute zero

____

____

SECTION SUMMARY

Read the following section summary.

- Temperature is a measure of the average kinetic energy of the particles of a substance.
- Fahrenheit, Celsius, and Kelvin are three temperature scales.
- Thermal expansion is the increase in volume of a substance due to an increase in temperature.
- Absolute zero (0 K, or –273°C) is the lowest possible temperature.
- A thermostat works because of the thermal expansion of a bimetallic strip.

Name ______________________ Class ______________ Date ____________

Skills Worksheet

Vocabulary and Section Summary

What Is Heat?

VOCABULARY

In your own words, write a definition of the following terms in the space provided.

1. heat

____

____

2. thermal energy

____

____

3. thermal conduction

____

____

4. thermal conductor

____

____

5. thermal insulator

____

____

6. convection

____

____

7. radiation

____

____

8. specific heat

____

____

SECTION SUMMARY

Read the following section summary.

- Heat is energy transferred between objects that are at different temperatures.
- Thermal energy is the total kinetic energy of the particles that make up a substance.
- Thermal energy will always be transferred from higher to lower temperature.
- Transfer of thermal energy ends when two objects that are in contact are at the same temperature.
- Conduction, convection, and radiation are three ways thermal energy is transferred.
- Specific heat is the amount of energy needed to change the temperature of 1 kg of a substance by 1°C.
- Energy transferred by heat cannot be measured directly. It must be calculated using specific heat, mass, and change in temperature.
- Energy transferred by heat is expressed in joules (J) and is calculated as follows: *heat* (J) = *specific heat* (J/kg•°C) × *mass* (kg) × *change in temperature* (°C).

Name ______________________ Class ______________ Date ____________

Skills Worksheet

Vocabulary and Section Summary

Matter and Heat

VOCABULARY

In your own words, write a definition of the following terms in the space provided.

1. states of matter

____

____

2. change of state

____

____

SECTION SUMMARY

Read the following section summary.

- States of matter include solid, liquid, and gas.
- Thermal energy transferred during a change of state does not change a substance's temperature. Rather, it causes a substance's particles to be rearranged.
- Chemical changes can cause thermal energy to be released or absorbed.
- A calorimeter can measure energy changes by measuring heat.

Name ______________________ Class ______________ Date ____________

Skills Worksheet

Vocabulary and Section Summary

Heat Technology

VOCABULARY

In your own words, write a definition of the following terms in the space provided.

1. insulation

____

____

2. heat engine

____

____

3. thermal pollution

____

____

SECTION SUMMARY

Read the following section summary.

- Central heating systems include hot-water heating systems and warm-air heating systems.
- Solar heating systems can be passive or active. In passive solar heating, a building takes advantage of the sun's energy without the use of moving parts. Active solar heating uses moving parts to aid the flow of solar energy throughout a building.
- Heat engines use heat to do work.
- The two kinds of heat engines are external combustion engines, which burn fuel outside the engine, and internal combustion engines, which burn fuel inside the engine.
- A cooling system transfers thermal energy from cooler temperatures to warmer temperatures by doing work.
- Transferring excess thermal energy to lakes and rivers can result in thermal pollution.

Name ______________________ Class ______________ Date ____________

Skills Worksheet

Directed Reading A

Section: Development of the Atomic Theory

THE BEGINNING OF ATOMIC THEORY

______ **1.** The word *atom* comes from the Greek word *atomos*, which means
- **a.** "dividable."
- **b.** "invisible."
- **c.** "hard particles."
- **d.** "not able to be divided."

______ **2.** Which of the following statements is a part of Democritus's theory about atoms?
- **a.** Atoms are small, soft particles.
- **b.** Atoms are always standing still.
- **c.** Atoms are made of a single material.
- **d.** Atoms are small particles that can be cut in half again and again.

3. We know that Democritus was right to say that all matter was made up of atoms. So why did people ignore Democritus's ideas for such a long time?

____

____

____

4. The smallest unit of an element that maintains the properties of that element is a(n) ______________________.

DALTON'S ATOMIC THEORY BASED ON EXPERIMENTS

______ **5.** Which of the following was NOT one of Dalton's theories?
- **a.** All substances are made of atoms.
- **b.** Atoms of the same element are exactly alike.
- **c.** Atoms of different elements are alike.
- **d.** Atoms join with other atoms to make new substances.

6. Dalton experimented with different substances. What did his results suggest?

____

____

____

THOMSON'S DISCOVERY OF ELECTRONS

7. In Thomson's experiments with a cathode-ray tube, he discovered that a(n) ______________________ charged plate attracted the beam. He concluded that the beam was made up of particles that have ______________________ electric charges.

8. The negatively charged subatomic particles that Thomson discovered are now called ______________________.

9. In Thomson's "plum-pudding" model, electrons are mixed throughout an ______________________.

RUTHERFORD'S ATOMIC "SHOOTING GALLERY"

______**10.** Before his experiment, what did Rutherford expect the particles to do?
- **a.** He expected the particles to pass right through the gold foil.
- **b.** He expected the particles to deflect to the sides of the gold foil.
- **c.** He expected the particles to bounce straight back.
- **d.** He expected the particles to become negatively charged.

11. What were the surprising results of Rutherford's gold-foil experiment?

____

____

____

WHERE ARE THE ELECTRONS?

______**12.** In 1911, Rutherford revised the atomic theory. Which of the following is NOT part of that theory?
- **a.** Most of the atom's mass is in its nucleus.
- **b.** The nucleus is a tiny, dense, positively charged region.
- **c.** Positively charged particles that pass close by the nucleus are pushed away by the positive charges in the nucleus.
- **d.** The nucleus is made up of protons and electrons.

13. The center of an atom is a dense region consisting of protons and neutrons called the ______________________.

14. What are electron clouds?

____

Name ______________________ Class ______________ Date ____________

Skills Worksheet

Directed Reading A

Section: The Atom

HOW SMALL IS AN ATOM?

______ **1.** Which of the following statements is true?
  a. A penny has about 20,000 atoms.
  b. A penny has more atoms than Earth has people.
  c. Aluminum is made up of large-sized atoms.
  d. Aluminum atoms have a diameter of about 3 cm.

WHAT IS AN ATOM MADE OF?

Match the correct description with the correct term. Write the letter in the space provided.

______ **2.** particle of the nucleus that has no electrical charge

______ **3.** particle found in the nucleus that is positively charged

______ **4.** particle with an unequal number of protons and electrons

______ **5.** negatively charged particle found outside the nucleus

______ **6.** contains most of the mass of an atom

______ **7.** SI unit used to express the masses of atomic particles

a. electron
b. atomic mass unit (amu)
c. nucleus
d. proton
e. ion
f. neutron

HOW DO ATOMS OF DIFFERENT ELEMENTS DIFFER?

8. The simplest atom is the ______________________ atom. It has one ______________________ and one ______________________.

9. Neutrons in the atom's ______________________ keep two or more protons from moving apart.

10. If you build an atom using two protons, two neutrons, and two electrons, you have built an atom of ______________________.

11. An atom does not have to have equal numbers of ______________________ and ______________________.

12. The number of protons in the nucleus of an atom is the

______________________ of that atom.

ISOTOPES

______ **13.** Isotopes always have
- **a.** the same number of protons.
- **b.** the same number of neutrons.
- **c.** a different atomic number.
- **d.** the same mass.

______ **14.** Which of the following is NOT true about unstable atoms?
- **a.** They are radioactive.
- **b.** They have a nucleus that always remains the same.
- **c.** They give off energy as they fall apart.
- **d.** They give off smaller particles as they fall apart.

______ **15.** What is the mass number of an isotope that has 5 protons, 6 neutrons, and 5 electrons?
- **a.** 1
- **b.** 11
- **c.** 10
- **d.** 16

______ **16.** If carbon has an atomic number of 6, how many neutrons does carbon-12 have?
- **a.** 12
- **b.** 8
- **c.** 6
- **d.** 18

17. Most elements contain a mixture of two or more ______________________.

18. The weighted average of the masses of all the naturally occurring isotopes of an element is the ______________________.

FORCES IN ATOMS

Match the correct definition with the correct term. Write the letter in the space provided.

______ **19.** helps protons stay together in the nucleus

______ **20.** pulls objects toward one another

______ **21.** an important force in radioactive atoms

______ **22.** holds the electrons around the nucleus

a. gravitational force

b. electromagnetic force

c. strong force

d. weak force

Name ______________________ Class ______________ Date __________

Skills Worksheet

Vocabulary and Section Summary

Development of the Atomic Theory

VOCABULARY

In your own words, write a definition of the following terms in the space provided.

1. atom

____

____

2. electron

____

____

3. nucleus

____

____

4. electron cloud

____

____

SECTION SUMMARY

Read the following section summary.

- Democritus thought that matter is composed of atoms.
- Dalton based his theory on observations of how elements combine.
- Thomson discovered electrons in atoms.
- Rutherford discovered that atoms are mostly empty space with a dense, positive nucleus.
- Bohr proposed that electrons are located in levels at certain distances from the nucleus.
- The electron-cloud model represents the current atomic theory.

Name ______________________ Class ______________ Date ____________

Skills Worksheet

Vocabulary and Section Summary

The Atom

VOCABULARY

In your own words, write a definition of the following terms in the space provided.

1. proton

____

____

2. atomic mass unit

____

____

3. neutron

____

____

4. atomic number

____

____

5. isotope

____

____

6. mass number

____

____

7. atomic mass

____

____

SECTION SUMMARY

Read the following section summary.

- Atoms are extremely small. Ordinary-sized objects are made up of very large numbers of atoms.
- Atoms consist of a nucleus, which has protons and usually neutrons, and electrons, located in electron clouds around the nucleus.
- The number of protons in the nucleus of an atom is that atom's atomic number. All atoms of an element have the same atomic number.
- Different isotopes of an element have different numbers of neutrons in their nuclei. Isotopes of an element share most chemical and physical properties.
- The mass number of an atom is the sum of the atom's neutrons and protons.
- Atomic mass is a weighted average of the masses of natural isotopes of an element.
- The forces at work in an atom are gravitational force, electromagnetic force, strong force, and weak force.

Name ______________________ Class ______________ Date ____________

Skills Worksheet

Directed Reading A

Section: Arranging the Elements

1. Why do you think scientists might have been frustrated by the organization of the elements before 1869?

____

____

____

____

____

____

DISCOVERING A PATTERN

______ 2. Which arrangement of elements did Mendeleev find produced a repeating pattern of properties?

a. by increasing density
b. by increasing melting point
c. by increasing shine
d. by increasing atomic mass

3. When something occurs or repeats at regular intervals, it is called

______________________.

4. Mendeleev's table, which shows elements' properties following a pattern that repeats every seven elements, is called the ______________________.

5. How was it possible that Mendeleev was able to predict the properties of elements that no one knew about?

____

____

____

____

____

____

CHANGING THE ARRANGEMENT

______ **6.** How did Moseley solve the problem of the elements that did not fit the pattern according to their properties?
- **a.** He rearranged the elements by atomic mass.
- **b.** He discovered protons, neutrons, and electrons.
- **c.** He disproved the periodic law.
- **d.** He determined the elements' atomic number and then arranged them by atomic number.

7. When the repeating chemical and physical properties of elements change periodically with the elements' atomic numbers, it is called the

______________________.

PERIODIC TABLE OF THE ELEMENTS

______ **8.** Which information is NOT included in each square of the periodic table in your text?
- **a.** atomic number
- **b.** chemical symbol
- **c.** melting point
- **d.** atomic mass

9. How can you tell on the periodic table that carbon is a solid at room temperature?

____

____

____

THE PERIODIC TABLE AND CLASSES OF ELEMENTS

10. Elements are classified as metals, nonmetals, or metalloids according to their

______________________.

11. The number of ______________________ in the outer energy level of an atom helps determine which category an element belongs in.

12. How can the zigzag line on the periodic table help you?

____

____

____

13. Most elements are ________________________, which can be found to the left of the zigzag line on the periodic table.

14. Most metals are ________________________, which means that they can be drawn into thin wires.

15. Most metals are ________________________ at room temperature.

16. Most metals are malleable. What does this mean?

____

____

17. What metal is flattened into sheets that are made into cans and foil?

____

18. What elements are found to the right of the zigzag line on the periodic table?

____

19. Semiconductors, also called ________________________, are the elements that border the zigzag line on the periodic table.

DECODING THE PERIODIC TABLE

______**20.** Which elements often share properties?

a. those in a period
b. those in a group
c. those with the same color
d. those in a horizontal row

______**21.** The physical and chemical properties of the elements change

a. within a group.
b. within a family.
c. across each period.
d. across each group.

22. For most elements, the ________________________ has one or two letters, with the first letter always capitalized.

23. Horizontal rows of elements on the periodic table are called

________________________.

24. Vertical columns of elements on the periodic table are called

________________________, or ________________________.

25. Some elements, such as ________________________, are named after scientists. Others, such as ________________________, are named after places.

Name ______________________ Class ______________ Date ____________

Skills Worksheet

Directed Reading A

Section: Grouping the Elements

______ **1.** What gives elements in a family or group similar properties?
- **a.** the same atomic mass
- **b.** the same number of protons in their nuclei
- **c.** the same number of electrons in their outer energy level
- **d.** the same number of total electrons

GROUP 1: ALKALI METALS

______ **2.** Which of the following is NOT true of alkali metals?
- **a.** They can be cut with a knife.
- **b.** They are usually stored in water.
- **c.** They are the most reactive of all the metals.
- **d.** They can easily give away their outer electron.

3. Elements in Group 1 of the periodic table are called ______________________.

GROUP 2: ALKALINE-EARTH METALS

4. Atoms of ______________________ have two outer-level electrons.

5. What are two products made from calcium compounds?

____

____

____

6. In what way does calcium help you?

____

____

____

7. Name three alkaline-earth metals besides calcium.

____

____

____

GROUPS 3–12: TRANSITION METALS

______ **8.** Which of the following characteristics does NOT describe transition metals?
- **a.** They are good conductors of thermal energy.
- **b.** They are more reactive than alkali and alkaline-earth metals.
- **c.** They have one or two electrons in the outer energy level.
- **d.** They are denser than elements in Groups 1 and 2.

9. Metals that are less reactive than alkali metals and alkaline-earth metals are called ______________________.

10. How is mercury different from other transition metals?

____

____

11. Two rows of transition metals are placed at the bottom of the periodic table to save space. Elements in the first row are called ______________________.

Elements in the second row are called ______________________.

12. Which lanthanide forms a compound that enables you to see red on a computer screen?

____

13. Which actinide is used in some smoke detectors?

____

GROUP 13: BORON GROUP

14. Why did Emperor Napoleon III of France use aluminum dinnerware?

____

____

____

15. What are some of the uses of aluminum?

____

____

____

GROUP 14: CARBON GROUP

16. The metalloids ______________________ and ______________________, both in Group 14, are used to make computer chips.

17. What are three compounds of carbon that are necessary for living things on Earth?

____

18. The hardest material known is ______________________.

19. What are some of the uses of diamond?

____

____

20. What form of carbon is used as a pigment?

____

GROUP 15: NITROGEN GROUP

21. Nitrogen is a ______________________ at room temperature.

22. Each element in the Nitrogen Group has ______________________ electrons in the outer level.

23. Nitrogen from the air can react with what element to make ammonia for fertilizer?

____

GROUP 16: OXYGEN GROUP

24. How is oxygen different from the other four elements in Group 16?

____

____

25. The element ______________________ can be found as a yellow solid in nature and is used to make sulfuric acid.

26. Why is oxygen important?

____

____

____

GROUP 17: HALOGENS

27. The atoms of ______________________ need to gain only one electron to have a complete outer level.

28. What important use do the halogens iodine and chlorine have in common?

____

____

29. Halogens combine with most metals to form ______________________, such as ______________________.

30. How does chlorinating water help protect people?

____

____

____

GROUP 18: NOBLE GASES

______ **31.** Which of the following statements about noble gases is NOT true?

a. They are colorless and odorless at room temperature.
b. They have a complete set of electrons in their outer energy level.
c. They normally react with other elements.
d. All of them are found in Earth's atmosphere in small amounts.

32. The atoms of ______________________ have a full set of electrons in their outer level.

33. The low ______________________ of helium makes blimps and weather balloons float.

HYDROGEN

______ **34.** Which of the following statements about hydrogen is NOT true?

a. It is useful as rocket fuel.
b. It is the most abundant element in the universe.
c. Its physical properties are closer to those of nonmetals than to those of metals.
d. It has two electrons in its outer energy level.

Name ______________________ Class ______________ Date ______________

Skills Worksheet

Vocabulary and Section Summary

Arranging the Elements

VOCABULARY

In your own words, write a definition of the following terms in the space provided.

1. periodic

____

____

2. periodic law

____

____

3. period

____

____

4. group

____

____

SECTION SUMMARY

Read the following section summary.

- Mendeleev developed the first periodic table by listing the elements in order of increasing atomic mass. He used his table to predict that elements with certain properties would be discovered later.
- Properties of elements repeat in a regular, or periodic, pattern.
- Moseley rearranged the elements in order of increasing atomic number.
- The periodic law states that the repeating chemical and physical properties of elements relate to and depend on elements' atomic numbers.
- Elements in the periodic table are classified as metals, nonmetals, and metalloids.
- Each element has a chemical symbol.
- A horizontal row of elements is called a *period.*
- Physical and chemical properties of elements change across each period.
- A vertical column of elements is called a *group* or *family.*
- Elements in a group usually have similar properties.

Name ______________________ Class ______________ Date ____________

Skills Worksheet

Vocabulary and Section Summary

Grouping the Elements

VOCABULARY

In your own words, write a definition of the following terms in the space provided.

1. alkali metal

____

____

2. alkaline-earth metal

____

____

3. halogen

____

____

4. noble gas

____

____

SECTION SUMMARY

Read the following section summary.

- Alkali metals (Group 1) are the most reactive metals. Atoms of the alkali metals have one electron in their outer level.
- Alkaline-earth metals (Group 2) are less reactive than the alkali metals are. Atoms of the alkaline-earth metals have two electrons in their outer level.
- Transition metals (Groups 3–12) include most of the well-known metals and the lanthanides and actinides.
- Groups 13–16 contain the metalloids and some metals and nonmetals.
- Halogens (Group 17) are very reactive nonmetals. Atoms of the halogens have seven electrons in their outer level.
- Noble gases (Group 18) are unreactive nonmetals. Atoms of the noble gases have a full set of electrons in their outer level.
- Hydrogen is set off by itself in the periodic table. Its properties do not match the properties of any one group.

Name ______________________ Class ______________ Date ____________

Skills Worksheet

Directed Reading A

Section: Electrons and Chemical Bonding

COMBINING ATOMS THROUGH CHEMICAL BONDING

______ 1. Which of the following substances results from combining atoms of carbon, hydrogen, and oxygen?

a. sugar
b. salt
c. water
d. sulfuric acid

______ 2. Which of the following is NOT true about electrons when chemical bonds form?

a. Electrons are shared.
b. Electrons are lost.
c. Electrons are destroyed.
d. Electrons are gained.

______ 3. Which of the following is an interaction that holds two atoms together?

a. chemical hold
b. chemical bond
c. chemical interaction
d. bond of chemicals

4. The joining of atoms to form new substances is called

______________________.

5. An explanation of a phenomenon that is based on observation, experimentation, and reasoning is a(n) ______________________.

6. People can use ______________________ to discuss theories of how and why atoms form bonds.

ELECTRON NUMBER AND ORGANIZATION

______ 7. How can you determine the number of electrons in an atom?

a. valence number
b. atomic number
c. chemical number
d. ionic number

______ 8. How many valence electrons are in an oxygen atom?

a. 2
b. 4
c. 6
d. 8

______ 9. What do elements within a group number have the same number of?

a. valance electrons
b. protons
c. neutrons
d. atoms

Match the correct description with the correct term. Write the letter in the space provided.

______ **10.** an electron in the outermost energy level — **a.** group

______ **11.** number of protons in an atom — **b.** valence electron

______ **12.** family on the periodic table to which an element belongs — **c.** atomic number

13. Which electrons in an atom make chemical bonds? Why?

____

____

14. How can the periodic table help you determine the number of valence electrons?

____

____

TO BOND OR NOT TO BOND

______ **15.** What determines whether an atom will form bonds?
- **a.** number of electrons
- **b.** number of valence electrons
- **c.** number of protons
- **d.** number of neutrons

______ **16.** Which group on the periodic table contains elements that do not normally form chemical bonds?
- **a.** Group 2
- **b.** Group 6
- **c.** Group 10
- **d.** Group 18

17. The outermost energy level of an atom is considered full if the level contains ______________________ electrons.

18. Helium atoms need only ______________________ valence electrons to have a filled outermost energy level.

19. The first energy level of any atom can hold only ______________________ electrons.

20. Why is it uncommon for noble gases to form chemical bonds?

____

____

21. Which is more likely to form bonds, an atom with 8 valence electrons or an atom with fewer than 8 valence electrons?

____

____

22. How can atoms with fewer than 8 valance electrons fill their outermost energy level? Use either sulfur or magnesium to explain the process.

____

____

____

____

____

____

Name ______________________ Class ______________ Date ____________

Skills Worksheet

Directed Reading A

Section: Ionic Bonds

FORMING IONIC BONDS

1. A chemical bond that forms when electrons are transferred from one atom to another is a(n) ______________________.

2. Charged particles that form when atoms gain or lose electrons are ______________________.

3. A transfer of electrons between atoms changes the number of electrons in an atom, but the number of ______________________ stays the same.

4. Why is an atom neutral?

____

____

____

5. Why are ions charged particles and thus no longer neutral?

____

____

____

____

FORMING POSITIVE IONS

_____ **6.** When atoms lose electrons through an ionic bond, they become
- **a.** positively charged.
- **b.** neutral.
- **c.** negatively charged.
- **d.** uncharged.

7. Most metals have few ______________________ and form positive ions.

8. If a sodium atom loses its only valence electron to another atom, the sodium atom becomes a sodium ______________________.

9. A sodium ion has a charge of ______________________.

10. The chemical symbol for a sodium ion is ______________________.

11. When electrons pull away from atoms, ______________________ is needed.

12. Where does the energy needed to take electrons from metals come from?

____

FORMING NEGATIVE IONS

______ **13.** Some atoms gain electrons during chemical changes and have a
- **a.** positive charge.
- **b.** negative charge.
- **c.** neutral charge.
- **d.** chemical charge.

______ **14.** The symbol for oxide is O^{2-}. How many electrons did the oxygen atom gain?
- **a.** 0
- **b.** 1
- **c.** 2
- **d.** 3

______ **15.** What ending is used for the names of negative ions?
- **a.** *-ion*
- **b.** *-ade*
- **c.** *-ide*
- **d.** *-ite*

16. Atoms of Group ____________________ elements give off the most energy when they gain an electron.

17. When is energy given off by most nonmetals?

____

____

18. When does an ionic bond form between a metal and a nonmetal?

____

____

____

IONIC COMPOUNDS

______ **19.** When ions bond, they form a repeating three-dimensional pattern called a(n)
- **a.** compound.
- **b.** crystal lattice.
- **c.** chemical bond.
- **d.** ionic bond.

20. Why does the compound formed by an ionic bond have a neutral charge when the ions that bond are charged?

____

____

____

21. List three properties of ionic compounds within a crystal lattice.

____

____

____

Name ______________________ Class ______________ Date ____________

Skills Worksheet

Directed Reading A

Section: Covalent and Metallic Bonds

COVALENT BONDS

______ **1.** What is formed when atoms share one or more pairs of electrons?

a. covalent bond
b. covalent compound
c. ionic bond
d. electric bond

______ **2.** What usually consists of two or more atoms joined in a definite ratio?

a. bond
b. valence electron
c. atom
d. molecule

3. A model that shows only the valence electrons in an atom is

a(n) electron-dot diagram.

COVALENT COMPOUNDS AND MOLECULES

4. What is the relationship between diatomic molecules and diatomic elements? Name one example of a diatomic element.

____

____

____

____

5. What is the smallest particle into which covalent bonds can be divided?

____

6. Name two examples of complex molecules.

____

METALLIC BONDS

7. A bond formed by the attraction between positively charged metal ions and the electrons in the metal is a(n) ______________________.

8. What allows valence electrons in metals to move throughout the metal?

____

____

____

PROPERTIES OF METALS

______ **9.** What property gives metals the ability to be drawn into wires?

a. malleability

b. conductivity

c. ductility

d. electricity

10. The property of ______________________ means that the metal can be hammered into sheets.

11. Give an example of how metallic bonding allows metals to conduct electric current.

____

____

____

____

12. Why doesn't a piece of metal break when it is bent?

____

____

____

Name ______________________________ Class ______________ Date ____________

Skills Worksheet

Vocabulary and Section Summary

Electrons and Chemical Bonding

VOCABULARY

In your own words, write a definition of the following terms in the space provided.

1. chemical bonding

___

___

2. chemical bond

___

___

3. valence electron

___

___

SECTION SUMMARY

Read the following section summary.

- Chemical bonding is the joining of atoms to form new substances. A chemical bond is an interaction that holds two atoms together.
- A valence electron is an electron in the outermost energy level of an atom.
- Most atoms form bonds by gaining, losing, or sharing electrons until they have 8 valence electrons. Atoms of some elements need only 2 electrons to fill their outermost level.

Name ______________________ Class ______________ Date ____________

Skills Worksheet

Vocabulary and Section Summary

Ionic Bonds

VOCABULARY

In your own words, write a definition of the following terms in the space provided.

1. ionic bond

____

____

2. ion

____

____

3. crystal lattice

____

____

SECTION SUMMARY

Read the following section summary.

- An ionic bond is a bond that forms when electrons are transferred from one atom to another. During ionic bonding, the atoms become oppositely charged ions.
- Ionic bonding usually occurs between atoms of metals and atoms of nonmetals.
- Energy is needed to remove electrons from metal atoms. Energy is released when most nonmetal atoms gain electrons.

Name ______________________ Class ______________ Date ____________

Skills Worksheet

Vocabulary and Section Summary

Covalent and Metallic Bonds

VOCABULARY

In your own words, write a definition of the following terms in the space provided.

1. covalent bond

____

____

2. molecule

____

____

3. metallic bond

____

____

SECTION SUMMARY

Read the following section summary.

- In covalent bonding, two atoms share electrons. A covalent bond forms when atoms share one or more pairs of electrons.
- Covalently bonded atoms form a particle called a *molecule*. A molecule is the smallest particle of a compound that has the chemical properties of the compound.
- In metallic bonding, the valence electrons move throughout the metal. A bond formed by the attraction between positive metal ions and the electrons in the metal is a metallic bond.
- Properties of metals include conductivity, ductility, and malleability.

Name ______________________ Class ______________ Date ____________

Skills Worksheet

Directed Reading A

Section: Forming New Substances

1. The substance that turns leaves green is called ______________________.

2. Why are leaves orange and yellow in the fall?

____

____

____

CHEMICAL REACTIONS

3. A process in which one or more substances change to form new substances is called a(n) ______________________.

4. How do the properties of the new substances compare with the properties of the original substances after a chemical change takes place?

____

____

____

5. A solid substance that is formed in a solution is called a(n) ______________________.

Match the correct example of a chemical reaction with the correct clue. Write the letter in the space provided.

______ **6.** Thermal energy is given off.

______ **7.** Precipitate forms.

______ **8.** Nitrogen dioxide forms.

______ **9.** Bleach spots form.

a. color change

b. energy change

c. solid formation

d. gas formation

10. When a gas is given off as a liquid boils, it is an example of a

______________________ change, not a ______________________ reaction.

11. What is the most important sign that a chemical reaction is occurring?

____

____

____

BONDS: HOLDING MOLECULES TOGETHER

12. What is a chemical bond?

____

____

____

13. How do new substances form during a chemical reaction?

____

____

____

____

14. What causes chemical bonds to break?

____

____

____

____

15. How many atoms make up a diatomic molecule?

____

16. What harmless substance forms from the reaction of sodium and chlorine gas?

____

Name ______________________ Class ______________ Date ____________

Skills Worksheet

Directed Reading A

Section: Chemical Formulas and Equations

CHEMICAL FORMULAS

______ **1.** About how many elements make up all known substances?
- **a.** 100
- **b.** 80
- **c.** 60
- **d.** 50

______ **2.** The subscript in the chemical formula H_2O tells you there are two
- **a.** atoms of hydrogen in the molecule.
- **b.** electrons on the hydrogen atom in the molecule.
- **c.** elements in the molecule.
- **d.** atoms of oxygen in the molecule.

______ **3.** What is the chemical formula for oxygen?
- **a.** O_2
- **b.** $C_6H_{12}O_6$
- **c.** H_2O
- **d.** $Ca(NO_3)_2$

______ **4.** What is the chemical formula for water?
- **a.** O_2
- **b.** $C_6H_{12}O_6$
- **c.** H_2O
- **d.** $Ca(NO_3)_2$

______ **5.** What is the chemical formula for glucose?
- **a.** O_2
- **b.** $C_6H_{12}O_6$
- **c.** H_2O
- **d.** $Ca(NO_3)_2$

6. A combination of chemical symbols and numbers that represent a substance is called a(n) ______________________.

7. What does a chemical formula show?

____

____

____

8. Covalent compounds are usually composed of two ______________________.

9. The formula for dinitrogen monoxide is ______________________.

10. The formula for carbon dioxide is ______________________.

11. Ionic compounds are composed of a(n) ______________________ and a(n) ______________________.

12. The overall charge of an ionic compound is ______________________.

Write the formula for each of the following ionic compounds.

13. sodium chloride ______________________________

14. magnesium chloride ______________________________

CHEMICAL EQUATIONS

15. What do musical notations and chemical equations have in common?

____

____

____

____

16. When chemical symbols and formulas are used as a shortcut to describe a chemical reaction, it is called a(n) ______________________.

17. A substance that forms in a chemical reaction is called a(n) ______________________.

18. A substance or molecule that participates in a chemical reaction is called a(n) ______________________.

19. When carbon reacts with oxygen to form carbon dioxide, carbon dioxide is the ______________________.

20. What will happen if the wrong chemical symbol or formula is used in a chemical equation?

____

____

____

____

21. In a chemical reaction, ______________________ are never gained or lost.

22. Antoine Lavoisier's work led to the ______________________.

23. What does the law of conservation of mass state?

____

____

24. A chemical equation must show the same numbers and kinds of

______________________ on both sides of the arrow.

25. The number placed in front of a chemical symbol or formula is called

a(n) ______________________.

26. How many oxygen atoms are contained in the formula $2H_2O$?

____

27. When balancing an equation, only ______________________ are changed,

not ______________________.

Name ______________________ Class ______________ Date ____________

Skills Worksheet

Directed Reading A

Section: Types of Chemical Reactions

______ **1.** Which of the following is NOT a type of chemical reaction?

a. synthesis
b. decomposition
c. single-displacement
d. double-decomposition

2. What do all four types of reactions have in common?

____

____

____

SYNTHESIS REACTIONS

3. When two or more substances combine to form one new compound, it is called a(n) ______________________.

DECOMPOSITION REACTIONS

4. When a single compound breaks down to form two or more simpler substances, it is called a(n) ______________________.

SINGLE-DISPLACEMENT REACTIONS

5. When an element replaces another element in a compound, it is called a(n) ______________________.

6. How is a person who cuts in on a dancing couple like a single-replacement reaction?

____

____

____

7. In a single-displacement reaction, a(n) ______________________ reactive element can replace a(n) ______________________ reactive element from a compound.

8. Most single-displacement reactions involve ______________________.

DOUBLE-DISPLACEMENT REACTIONS

9. In a double-displacement reaction, a(n) ______________________, a(n) ______________________, or a(n) ______________________ forms from the exchange of ions between two compounds.

Match the correct description with the correct term. Write the letter in the space provided.

_____ **10.** Zinc reacts with hydrochloric acid to form zinc chloride and hydrogen.

_____ **11.** Carbonic acid decomposes to form water and carbon dioxide.

_____ **12.** Sodium reacts with chlorine to form sodium chloride.

_____ **13.** Sodium fluoride and silver chloride are formed from the reaction of sodium chloride with silver fluoride.

a. decomposition

b. double-displacement

c. single-displacement

d. synthesis

Name ______________________ Class ______________ Date ______________

Skills Worksheet

Directed Reading A

Section: Energy and Rates of Chemical Reactions

1. All chemical reactions either give off or absorb ______________.

REACTIONS AND ENERGY

2. Why is chemical energy needed in all chemical reactions?

____

____

3. When energy is released during a chemical reaction, it is called a(n)

______________ reaction.

4. Give one example of a type of energy released in exothermic reactions.

____

____

5. When energy is taken in during a chemical reaction, it is called

a(n) ______________ reaction.

6. Photosynthesis is an example of a(n) ______________ process.

7. What does the law of conservation of energy state?

____

____

____

8. What two things can happen to energy in a chemical reaction?

____

____

9. What happens to the energy taken in during endothermic reactions?

____

____

RATES OF REACTIONS

10. The speed at which new particles form is called

the ______________________.

11. The smallest amount of energy needed to start a chemical reaction is

called ______________________.

12. Name three sources of activation energy.

____

____

____

FACTORS AFFECTING RATES OF REACTIONS

13. What four factors affect how rapidly a chemical reaction takes place?

____

____

____

14. As temperature increases, the rate of reaction gets ______________________.

15. A measure of the amount of one substance that is dissolved in another is

called ______________________.

16. How does a high concentration of reactants affect the rate of reaction?

____

____

____

17. The amount of exposed surface of a substance is called

______________________.

18. How can you increase the surface area of a solid reactant?

____

____

____

19. A substance that slows down or stops a chemical reaction is called

a(n) ______________________.

20. Give two examples of an inhibitor.

____

____

21. A substance that speeds up a reaction without being permanently changed is called a(n) ______________________.

22. How does a catalyst enable a reaction to happen more quickly?

____

____

____

Name ______________________ Class ______________ Date ______________

Skills Worksheet

Vocabulary and Section Summary

Forming New Substances

VOCABULARY

In your own words, write a definition of the following terms in the space provided.

1. chemical reaction

____

____

2. precipitate

____

____

SECTION SUMMARY

Read the following section summary.

- A chemical reaction is a process by which substances change to produce new substances with new chemical and physical properties.
- Signs that indicate a chemical reaction has taken place are a color change, formation of a gas or a solid, and release of energy.
- During a reaction, bonds are broken, atoms are rearranged, and new bonds are formed.

Name ______________________ Class ______________ Date ____________

Skills Worksheet

Vocabulary and Section Summary

Chemical Formulas and Equations

VOCABULARY

In your own words, write a definition of the following terms in the space provided.

1. chemical formula

____

____

2. chemical equation

____

____

3. reactant

____

____

4. product

____

____

5. law of conservation of mass

____

____

SECTION SUMMARY

Read the following section summary.

- A chemical formula uses symbols and subscripts to describe the makeup of a compound.
- Chemical formulas can often be written from the names of covalent and ionic compounds.
- A chemical equation uses chemical formulas, chemical symbols, and coefficients to describe a reaction.
- Balancing an equation requires that the same numbers and kinds of atoms be on each side of the equation.
- A balanced equation illustrates the law of conservation of mass: mass is neither created nor destroyed during ordinary physical and chemical changes.

Name ______________________ Class ______________ Date ____________

Skills Worksheet

Vocabulary and Section Summary

Types of Chemical Reactions

VOCABULARY

In your own words, write a definition of the following terms in the space provided.

1. synthesis reaction

____

____

2. decomposition reaction

____

____

3. single-displacement reaction

____

____

4. double-displacement reaction

____

____

SECTION SUMMARY

Read the following section summary.

- A synthesis reaction is a reaction in which two or more substances combine to form a compound.
- A decomposition reaction is a reaction in which a compound breaks down to form two or more simpler substances.
- A single-displacement reaction is a reaction in which an element takes the place of another element that is part of a compound.
- A double-displacement reaction is a reaction in which ions in two compounds exchange places.

Name ______________________ Class ______________ Date ____________

Skills Worksheet

Vocabulary and Section Summary

Energy and Rates of Chemical Reactions

VOCABULARY

In your own words, write a definition of the following terms in the space provided.

1. exothermic reaction

____

____

2. endothermic reaction

____

____

3. law of conservation of energy

____

____

4. activation energy

____

____

5. inhibitor

____

____

6. catalyst

____

____

SECTION SUMMARY

Read the following section summary.

- Energy is given off in exothermic reactions.
- Energy is absorbed in an endothermic reaction.
- The law of conservation of energy states that energy is neither created nor destroyed.
- Activation energy is the energy needed for a reaction to occur.
- The rate of a chemical reaction is affected by temperature, concentration, surface area, and the presence of an inhibitor or catalyst.

Name ______________________ Class ______________ Date ____________

Skills Worksheet

Directed Reading A

Section: Ionic and Covalent Compounds

______ **1.** What is the force of attraction that holds atoms or ions together called?

a. valence electrons
b. ionic compounds
c. chemical bond
d. compound cement

______ **2.** What are the electrons found in the outermost energy levels of an atom called?

a. valence electrons
b. ionic electrons
c. covalent electrons
d. compound electrons

IONIC COMPOUNDS AND THEIR PROPERTIES

______ **3.** An ionic bond is an attraction between

a. positively charged ions.
b. oppositely charged ions.
c. negatively charged ions.
d. metallic ions.

______ **4.** When a metal meets a nonmetal, electrons are transferred and the metal atoms become

a. positively charged.
b. neutral.
c. negatively charged.
d. oppositely charged.

______ **5.** When a metal meets a nonmetal, the nonmetal atom becomes

a. positively charged.
b. neutral.
c. negatively charged.
d. oppositely charged.

______ **6.** Table salt is formed when an electron is transferred from a sodium atom to a

a. metal atom.
b. chlorine atom.
c. nonmetal atom.
d. positively charged atom.

______ **7.** Ionic compounds tend to be brittle solids

a. at room temperature.
b. at high temperatures.
c. outdoors.
d. when wet.

______ **8.** In a crystal lattice each ion is bonded to the

a. pattern it is made with.
b. ions around it.
c. compound around it.
d. crystal's edge.

______ **9.** When an ionic compound is hit, the pattern shifts, ions repel each other, and the crystal

a. becomes more solid.
b. forms a new lattice.
c. breaks apart.
d. becomes bonded.

______**10.** Because strong ionic bonds hold ions together, ionic compounds have
a. a low melting point.
b. a lukewarm melting point.
c. a high melting point.
d. a variable melting point.

______**11.** Many ionic compounds dissolve easily
a. in air.
b. at high temperatures.
c. in water.
d. in electric current.

12. When an ionic compound dissolves in water, why can it conduct electric current?

____

____

COVALENT COMPOUNDS AND THEIR PROPERTIES

______**13.** Covalent compounds are formed when a group of atoms share
a. uncharged particles.
b. neutrons.
c. protons.
d. electrons.

______**14.** Compared with ionic bonds, covalent bonds are
a. weaker.
b. stronger.
c. larger.
d. smaller.

______**15.** The group of atoms that make up a covalent compound is called a(n)
a. bond.
b. electron.
c. molecule.
d. atom.

16. What does it mean if a substance is not soluble in water?

____

____

17. Why are covalent compounds often not soluble in water?

____

____

18. Why do covalent compounds have lower melting points than ionic compounds?

____

____

____

____

19. Why doesn't sugar dissolved in water conduct electric current?

____

____

____

____

20. How are acids that have been dissolved in water able to conduct an electric current?

____

____

____

Name ______________________ Class ______________ Date ____________

Skills Worksheet

Directed Reading A

Section: Acids and Bases

ACIDS AND THEIR PROPERTIES

______ **1.** What is any compound that increases the number of hydronium (H_3O^+) ions dissolved in water called?
a. base
b. acid
c. indicator
d. neutral

______ **2.** What does each hydrogen ion bond with to form hydronium ions?
a. an oxygen particle
b. a water molecule
c. an acid
d. tea

______ **3.** What do hydrogen ions, H^+, form when they bond to water molecules, H_2O?
a. hydrogen ions, H^+
b. hydronium ions, H_3O^+
c. water molecules, H_2O
d. bases

______ **4.** What flavor do acids have?
a. sweet
b. salty
c. sour
d. crunchy

______ **5.** Why should a person NEVER taste or touch an unknown chemical?
a. many are flavorless
b. many are too sweet
c. many are corrosive
d. many are too salty

______ **6.** What can corrosive substances destroy?
a. sour things
b. poisons
c. lemons
d. body tissues and clothing

______ **7.** A compound that can reversibly change color depending on conditions such as pH is called a(n)
a. indicator.
b. color meter.
c. color changer.
d. water molecule.

______ **8.** Two commonly used indicators are bromthymol blue and
a. hydrochloric acid.
b. silver nitrate.
c. litmus paper.
d. color changer.

______ **9.** What color does blue litmus paper turn when acid is added to it?
a. green
b. red
c. blue
d. orange

______ **10.** What is produced when acids react with some metals?
a. oxygen gas
b. metals
c. silver crystals
d. hydrogen gas

______ **11.** Since acids form hydronium ions in water, solutions of acids can
a. make oxygen.
b. break apart water molecules.
c. conduct electric current.
d. straighten hair.

Match the correct acid with the product in which it is used. Write the letter in the space provided.

______ **12.** rubber

______ **13.** car batteries

______ **14.** orange juice

______ **15.** swimming pools

______ **16.** soft drinks

a. sulfuric acid
b. nitric acid
c. hydrochloric acid
d. citric acid
e. carbonic acid

BASES AND THEIR PROPERTIES

______ **17.** Any compound that increases the number of hydroxide ions when dissolved in water is a(n)
a. gas.
b. sodium.
c. acid.
d. base.

______ **18.** The properties of bases include a bitter taste and a(n)
a. strong bond.
b. slippery feel.
c. hydroxide lattice.
d. unpleasant odor.

______ **19.** What should you NEVER do to identify a chemical?
a. add salt to it
b. use an indicator
c. taste or touch it
d. look in a book

______ **20.** What color does a base change red litmus paper to?
a. blue
b. purple
c. green
d. orange

______ **21.** Because bases increase the number of hydroxide ions, OH^-, solutions of bases can
a. indicate temperature.
b. split atoms.
c. conduct electric current.
d. stop electric current.

Match the correct base with the product in which it is used. Write the letter in the space provided.

______ **22.** soap

______ **23.** antacids

______ **24.** cement

a. magnesium hydroxide
b. sodium hydroxide
c. calcium hydroxide

Name ______________________ Class ______________ Date ______________

Skills Worksheet

Directed Reading A

Section: Solutions of Acids and Bases

STRENGTHS OF ACIDS AND BASES

______ **1.** What is the amount of acid or base dissolved in water called?
a. concentration
b. strength
c. pH
d. neutralization

______ **2.** When an acid or base dissolves in water, what is dependent on the number of molecules that break apart?
a. its concentration
b. its weakness
c. its durability
d. its strength

______ **3.** In what kind of solution do all the molecules of an acid break apart in water?
a. a strong acid
b. a strong base
c. a weak acid
d. a weak base

______ **4.** In what kind of solution do only a few of the molecules of an acid break apart in water?
a. a strong acid
b. a strong base
c. a weak acid
d. a weak base

______ **5.** In what kind of solution do all the molecules of a base break apart?
a. a strong acid
b. a strong base
c. a weak acid
d. a weak base

______ **6.** What is a solution called when only a few molecules of a base break apart?
a. a strong acid
b. a strong base
c. a weak acid
d. a weak base

ACIDS, BASES, AND NEUTRALIZATION

______ **7.** What is the reaction between acids and bases called?
a. neutralization reaction
b. explosion
c. strength
d. evaporation

______ **8.** What do the H^+ ions of an acid and the OH^- ions of a base form when they react?
a. oxygen
b. water
c. sugar
d. hydrogen gas

______ **9.** What can show whether a solution contains an acid or a base?
a. an indicator
b. pure water
c. antacids
d. salt

10. A value that is used to express the acidity or basicity (alkalinity) of a system is called ______________________.

11. The pH of a solution shows the concentration of what type of ion?

____

____

12. What is the pH of a neutral solution?

____

____

13. What type of solution has a pH greater than 7?

____

____

14. What type of solution has a pH less than 7?

____

____

15. What are three examples of common materials with a pH of less than 7?

____

____

____

16. What are three examples of common materials with a pH of more than 7?

____

____

____

For each organism listed, write the preferred pH or pH range.

______________________ **17.** pine trees

______________________ **18.** lettuce

______________________ **19.** fish

20. How does acid rain form, and what is its effect on nature?

____

____

____

SALTS

21. What two substances are produced when an acid neutralizes a base?

____

____

22. What is a salt, and how does it form?

____

____

23. Name two salts, and tell what they are used for.

____

____

____

____

Name ______________________ Class ______________ Date ____________

Skills Worksheet

Directed Reading A

Section: Organic Compounds

______ **1.** What is a covalent compound composed of carbon-based molecules called?
a. hydrogen atom
b. oxygen atom
c. organic compound
d. valence electron

THE FOUR BONDS OF A CARBON ATOM

______ **2.** How many valence electrons does each carbon atom have?
a. three
b. two
c. six
d. four

______ **3.** What do structural formulas show about the atoms in a molecule?
a. what colors the atoms are
b. how the atoms are connected
c. how heavy the atoms are
d. what size the atoms are

______ **4.** What do the backbones of some compounds have hundreds or thousands of?
a. carbon atoms
b. carbon molecules
c. structural formulas
d. acid ions

HYDROCARBONS AND OTHER ORGANIC COMPOUNDS

______ **5.** What is an organic compound composed only of carbon and hydrogen called?
a. molecule
b. electron
c. hydrocarbon
d. single bond

______ **6.** What is a hydrocarbon in which each carbon atom in the molecule shares a single bond with each of the four other atoms called?
a. unsaturated hydrocarbon
b. saturated hydrocarbon
c. bonded hydrocarbon
d. unbonded hydrocarbon

______ **7.** What is another name for a saturated hydrocarbon?
a. carbon atom
b. alkane
c. triple bond
d. atomic bond

______ **8.** What is a hydrocarbon in which at least one pair of carbon atoms share a double or triple bond called?
a. unsaturated hydrocarbon
b. saturated hydrocarbon
c. bonded hydrocarbon
d. unbonded hydrocarbon

______ **9.** What are compounds that contain two carbon atoms connected by a double bond called?
a. alkanes
b. double-binds
c. alkenes
d. alkynes

______ **10.** What are compounds that contain two carbon atoms connected by a triple bond called?
a. alkanes
b. triple-binds
c. alkenes
d. alkynes

11. What are aromatic compounds usually based on?

____

____

12. What kind of bonds do the atoms in a ring of benzene have?

____

____

13. What do aromatic hydrocarbons often have?

____

____

14. List three elements that other organic compounds might have in them.

____

____

BIOCHEMICALS: THE COMPOUNDS OF LIFE

______ **15.** Carbohydrates, lipids, proteins, and nucleic acids are the four categories of
a. living things.
b. unsaturated hydrocarbons.
c. organic compounds.
d. biochemicals.

______ **16.** Carbohydrates are biochemicals that are composed of one or more
a. saturated hydrocarbons.
b. sugar molecules.
c. organic compounds.
d. starch molecules.

______ **17.** Carbohydrates are used as a source of
a. fat.
b. genetic material.
c. energy.
d. structure.

______ **18.** Simple carbohydrates are made up of
a. simple sugars.
b. cellulose.
c. proteins.
d. lipids.

______ **19.** Complex carbohydrates are made of hundreds or thousands of
a. lipids.
b. sugar molecules.
c. proteins.
d. nucleic acids.

______ **20.** Lipids are biochemicals that do not
a. store excess energy.
b. make up cell membranes.
c. dissolve in water.
d. store vitamins.

______ **21.** Proteins are biochemicals made up of "building blocks" called
a. sugars.
b. amino acids.
c. nucleic acids.
d. lipids.

______ **22.** If a single amino acid is missing or out of place, the protein
a. may not include sulfur.
b. may not provide support.
c. may not transport materials.
d. may not function correctly.

23. List three roles that proteins have in your body and in other living things.

____

____

24. What are the largest molecules made by living organisms called?

____

____

25. What are nucleic acids made up of?

____

____

26. What is the only reason living things differ from each other?

____

____

27. Since nucleic acids contain all the information needed for a cell to make its proteins, what are nucleic acids sometimes called?

____

____

28. What are the two kinds of nucleic acids, and what are their functions?

____

____

Name ______________________ Class ______________ Date ____________

Skills Worksheet

Vocabulary and Section Summary

Ionic and Covalent Compounds

VOCABULARY

In your own words, write a definition of the following terms in the space provided.

1. chemical bond

____

____

2. ionic compound

____

____

3. covalent compound

____

____

SECTION SUMMARY

Read the following section summary.

- Ionic compounds have ionic bonds between ions of opposite charges.
- Ionic compounds are usually brittle, have high melting points, dissolve in water, and often conduct an electric current.
- Covalent compounds have covalent bonds and consist of particles called *molecules*.
- Covalent compounds have low melting points, don't dissolve easily in water, and do not conduct electric current.

Name ______________________ Class ______________ Date ____________

Skills Worksheet

Vocabulary and Section Summary

Acids and Bases

VOCABULARY

In your own words, write a definition of the following terms in the space provided.

1. acid

____

____

2. indicator

____

____

3. base

____

____

SECTION SUMMARY

Read the following section summary.

- An acid is a compound that increases the number of hydronium ions in solution.
- Acids taste sour, turn blue litmus paper red, react with metals to produce hydrogen gas, and may conduct an electric current when in solution.
- Acids are used for industrial purposes and in household products.
- A base is a compound that increases the number of hydroxide ions in solution.
- Bases taste bitter, feel slippery, and turn red litmus paper blue. Most solutions of bases conduct an electric current.
- Bases are used in cleaning products and acid neutralizers.

Name ______________________ Class ______________ Date ____________

Skills Worksheet

Vocabulary and Section Summary

Solutions of Acids and Bases

VOCABULARY

In your own words, write a definition of the following terms in the space provided.

1. neutralization reaction

____

____

2. pH

____

____

3. salt

____

____

SECTION SUMMARY

Read the following section summary.

- Every molecule of a strong acid or base breaks apart to form ions. Few molecules of weak acids and bases break apart to form ions.
- An acid and a base can neutralize one another to make salt and water.
- pH is a measure of hydronium ion concentration in a solution.
- A salt is an ionic compound formed in a neutralization reaction. Salts have many industrial and household uses.

Name ______________________ Class ______________ Date ____________

Skills Worksheet

Vocabulary and Section Summary

Organic Compounds

VOCABULARY

In your own words, write a definition of the following terms in the space provided.

1. organic compound

____

____

2. hydrocarbon

____

____

3. carbohydrate

____

____

4. lipid

____

____

5. protein

____

____

6. nucleic acid

____

____

SECTION SUMMARY

Read the following section summary.

- Organic compounds contain carbon, which can form four bonds.
- Hydrocarbons are composed of only carbon and hydrogen.
- Hydrocarbons may be saturated, unsaturated, or aromatic hydrocarbons.
- Carbohydrates are made of simple sugars.
- Lipids store energy and make up cell membranes.
- Proteins are composed of amino acids.
- Nucleic acids store genetic information and help cells make proteins.

Name ______________________ Class ______________ Date ______________

Skills Worksheet

Directed Reading A

Section: Radioactivity

DISCOVERING RADIOACTIVITY

______ **1.** What happens to fluorescent minerals when light shines on them?
a. The minerals explode.
b. The minerals break into particles.
c. The minerals glow.
d. The minerals give off gases.

______ **2.** Becquerel made a hypothesis that fluorescent minerals give off
a. minerals.
b. X rays.
c. uranium.
d. particles.

______ **3.** In his experiment, Becquerel discovered that a fluorescent mineral made an image on a photographic plate even though there was no
a. energy.
b. uranium.
c. light.
d. X rays.

4. Becquerel concluded that the energy that made the image on the plate came from an element called ______________________.

5. Energy in the form of particles and rays emitted by the nuclei of some atoms is called ______________________.

6. Marie Curie named the process Becquerel discovered ______________________ or radioactive decay.

KINDS OF RADIOACTIVE DECAY

______ **7.** What does the unstable nucleus of an atom give off during radioactive decay?
a. particles and energy
b. molecules and energy
c. particles and gases
d. molecules and gases

Match the correct description with the correct term. Write the letter in the space provided.

_____ **8.** the release of an electron or a positron from the nucleus of an atom

_____ **9.** the release of gamma rays from the nucleus of an atom

_____ **10.** the release of a particle composed of two protons and two neutrons from the nucleus of an atom

a. gamma decay

b. alpha decay

c. beta decay

11. Particles released during alpha decay are called ____________________.

12. The sum of the numbers of protons and neutrons in the nucleus of an atom is called the ____________________.

13. Particles released during ____________________ are made up of two protons and two neutrons.

14. Mass number and charge are conserved in ____________________.

15. An electron or positron released during beta decay is called a(n) ____________________.

16. Explain why the mass number of a beta particle is 0.

____

____

17. Describe what occurs when a carbon-14 nucleus undergoes beta decay.

____

____

____

18. Atoms of an element that have the same number of protons as other atoms of that element, but a different number of neutrons are called ____________________.

19. Explain what happens during any beta decay.

____

____

____

20. The release of gamma rays from a nucleus is

called ______________________.

21. Does gamma decay by itself cause one element to become another element, as in alpha decay and beta decay? Explain.

____

____

____

THE PENETRATING POWER OF RADIATION

______ **22.** What can happen when matter is hit by nuclear radiation?

a. An atom can give up protons and electrons.
b. An atom can give up protons, and chemical bonds can break.
c. An atom can give up positrons, and chemical bonds can break.
d. An atom can give up electrons, and chemical bonds can break.

23. A single large exposure to radiation can cause ______________________ in a person.

24. List three of the symptoms of radiation sickness.

____

____

____

Match the correct description with the correct term. Write the letter in the space provided.

______ **25.** can be stopped by 3 mm of aluminum

______ **26.** can be stopped by paper or clothing

______ **27.** can be stopped by a few centimeters of lead or a few meters of concrete

a. beta particles
b. gamma rays
c. alpha particles

28. Give an example of how radiation can damage nonliving matter.

____

____

29. Which of the three types of nuclear radiation does the most damage from inside living things? Explain why.

____

____

____

FINDING A DATE BY DECAY

______ **30.** What is the half-life of carbon-14?
- **a.** 6,750 years
- **b.** 5,300 years
- **c.** 5,750 years
- **d.** 5,730 years

______ **31.** What part of an original sample remains after three half-lives?
- **a.** three-eighths
- **b.** one-eighth
- **c.** three-fourths
- **d.** one-fourth

32. The time required for half of the nuclei of a radioactive isotope to decay is called its ______________________.

33. Carbon-14 can help determine the age of objects up to 50,000 years old. How do scientists find the age of dinosaur fossils?

____

____

____

USES OF RADIOACTIVITY

34. What is one medical use for radioactive materials?

____

____

35. What is one way radioactive isotopes can be used to detect defects in metal structures?

____

____

Name ______________________ Class ______________ Date ______________

Skills Worksheet

Directed Reading A

Section: Energy from the Nucleus

NUCLEAR FISSION

______ **1.** The process by which a large nucleus splits into two smaller nuclei and releases energy is called
- **a.** nuclear fusion.
- **b.** nuclear fission.
- **c.** neutronic fission.
- **d.** neutronic fusion.

______ **2.** How can large atoms be forced to undergo nuclear fission?
- **a.** by hitting the atoms with protons
- **b.** by hitting the atoms with electrons
- **c.** by hitting the atoms with neutrons
- **d.** by hitting the neutrons with atoms

______ **3.** The nuclear fission of the uranium in one fuel pellet releases as much energy as burning
- **a.** 100 kg of coal
- **b.** 1,000 kg of coal
- **c.** 10,000 kg of coal
- **d.** 10 kg of coal

4. In nuclear fission, the total mass of products is less than the mass of the original reactants. How can you explain this?

____

____

____

5. A continuous series of nuclear fission reactions is called a(n) ______________.

6. Describe what happens during an uncontrolled nuclear reaction.

____

____

____

____

7. Is the nuclear reaction in a nuclear power plant controlled or uncontrolled? Explain.

____

____

____

Put the steps in the process of generating electricity in a nuclear power plant in order from 1 to 5. Write the appropriate number on the space provided.

______ **8.** Energy is absorbed by a coolant.

______ **9.** A generator changes mechanical energy into electrical energy.

______ **10.** Uranium-235 nuclei undergo a nuclear chain reaction.

______ **11.** Water changes to steam as it absorbs energy from hot coolant.

______ **12.** Steam turns a turbine attached to a generator.

ADVANTAGES AND DISADVANTAGES OF FISSION

______ **13.** What are the two main problems with using nuclear fission to generate electrical energy?

a. the risk of losing power, and storing waste
b. the risk of accidents, and storing power
c. the risk of war, and storing waste
d. the risk of accidents, and storing waste

______ **14.** What happened during the 1986 nuclear disaster at Chernobyl, Ukraine?

a. Uranium ignited, sending waste into the atmosphere.
b. The reactor exploded, sending waste into the atmosphere.
c. Uncontrolled nuclear fusion sent waste into the atmosphere.
d. Emergency systems failed, sending waste into the atmosphere.

15. How is the waste created by nuclear power plants in the United States stored?

____

____

____

16. What are three advantages and three disadvantages of nuclear power plants compared with power plants that burn fossil fuels?

____

____

____

NUCLEAR FUSION

17. When two or more nuclei that have small masses combine, or fuse, to form a larger nucleus, it is called ______________________.

18. Matter in which electrons have been removed from atoms is called ______________________.

19. Explain what must happen for nuclear fusion to occur.

____

____

____

____

20. What is one place with temperatures high enough for nuclear fusion to happen?

____

____

ADVANTAGES AND DISADVANTAGES OF FUSION

21. What is one problem with generating power by nuclear fusion?

____

____

22. Would an explosion of a fusion reactor be more harmful or less harmful than an explosion of a fission reactor? Explain.

____

____

____

____

Name ______________________ Class ______________ Date ____________

Skills Worksheet

Vocabulary and Section Summary

Radioactivity

VOCABULARY

In your own words, write a definition of the following terms in the space provided.

1. radioactivity

____

____

2. mass number

____

____

3. isotope

____

____

4. half-life

____

____

SECTION SUMMARY

Read the following section summary.

- Henri Becquerel discovered radioactivity while trying to study X rays. Radioactivity is the process by which a nucleus gives off nuclear radiation.
- An alpha particle is composed of two protons and two neutrons. A beta particle can be an electron or a positron. Gamma rays are a form of light with very high energy.
- Gamma rays penetrate matter better than alpha or beta particles do. Beta particles penetrate matter better than alpha particles do.
- Nuclear radiation can damage living and nonliving matter.
- Half-life is the amount of time it takes for one-half of the nuclei of a radioactive isotope to decay. The age of some objects can be determined using half-lives.
- Uses of radioactive materials include detecting defects in materials, sterilizing products, diagnosing illness, and generating electrical energy.

Name ______________________ Class ______________ Date ____________

Skills Worksheet

Vocabulary and Section Summary

Energy from the Nucleus

VOCABULARY

In your own words, write a definition of the following terms in the space provided.

1. nuclear fission

____

____

2. nuclear chain reaction

____

____

3. nuclear fusion

____

____

SECTION SUMMARY

Read the following section summary.

- In nuclear fission, a massive nucleus breaks into two nuclei.
- In nuclear fusion, two or more nuclei combine to form a larger nucleus.
- Nuclear fission is used in power plants to generate electrical energy. A limited fuel supply and radioactive waste products are disadvantages of fission.
- Nuclear fusion cannot yet be used as an energy source, but plentiful fuel and little waste are advantages of fusion.

Name ______________________ Class ______________ Date ____________

Skills Worksheet

Directed Reading A

Section: Electric Charge and Static Electricity

ELECTRIC CHARGE

______ **1.** What do you call the tiny particles that make up matter?
- **a.** electricity
- **b.** atoms
- **c.** electrons
- **d.** charges

______ **2.** Atoms are made up of protons, neutrons, and what third particle?
- **a.** charges
- **b.** electricity
- **c.** electrons
- **d.** forces

3. What three types of charge can an object have?

____

____

____

4. What does the law of electric charges state?

____

____

5. Protons are ______________________ charged.

6. Electrons are ______________________ charged.

7. The force between charged objects is a(n) ______________________.

8. What two things affect the size of the electric force?

____

____

9. The region around a charged object where an electric force is exerted on another charged object is the ______________________.

10. How do charged objects within an electric field interact?

____

CHARGE IT!

11. Why are atoms uncharged?

____

12. What happens when an object loses electrons?

____

13. What happens when an object gains electrons?

____

Match the correct definition with the correct term. Write the letter in the space provided.

_____**14.** This happens when electrons are "wiped" from one object to another.

_____**15.** This happens when charges in an uncharged metal object are rearranged without direct contact with a charged object.

_____**16.** This happens when electrons move by direct contact.

a. conduction

b. induction

c. friction

17. When you charge objects by any method, no charges are

____________________ or ____________________.

18. You can use a device called a(n) ____________________ to see if something is charged.

19. Can you tell if an object has a positive or negative charge with an electroscope?

____

MOVING CHARGES

_____**20.** Which of the following is a material in which charges can move easily?

a. electrical conductor

b. electrical insulator

c. electrical jumper

d. electrical stopper

______**21.** Which of the following is a material in which charges CANNOT move easily?
a. electrical conductor
b. electrical insulator
c. electrical jumper
d. electrical stopper

______**22.** Which of the following is a good conductor?
a. wood
b. air
c. copper
d. glass

______**23.** Which of the following is a good insulator?
a. aluminum
b. glass
c. mercury
d. copper

24. Why are most metals good conductors?

____

____

25. What factors make a material a good insulator?

____

____

STATIC ELECTRICITY

26. What is static electricity?

____

____

27. The loss of static electricity as charges move off an object is called

______________________.

28. What three things might you notice after an electric discharge?

____

____

____

29. How does lightning occur within a cloud?

____

____

____

30. Why is it unsafe to be at the beach during a lightning storm?

____

____

31. How do lightning rods protect buildings from lightning?

____

____

Name ______________________ Class ______________ Date ____________

Skills Worksheet

Directed Reading A

Section: Electric Current and Electrical Energy

1. The energy of electric charges is called ________________________.

ELECTRIC CURRENT

_____ **2.** What is the rate at which charges pass a given point?
- **a.** electrical energy
- **b.** amps
- **c.** electric current
- **d.** voltage

_____ **3.** Which of the following is the unit for electric current?
- **a.** amperes, or amps
- **b.** ohms
- **c.** volts
- **d.** charges

_____ **4.** Which of these letters is used to represent current in equations?
- **a.** C
- **b.** I
- **c.** R
- **d.** T

_____ **5.** What are the two types of electric current?
- **a.** electrons and neutrons
- **b.** alternating and direct
- **c.** current and electric
- **d.** charge and negative

6. When you flip the switch on a flashlight, a(n) ________________________ is set up in the wire.

7. When charges continually shift from flowing in one direction to flowing in the reverse direction, there is a(n) ________________________ current.

8. When charges always flow in the same direction, there is a(n) ________________________ current.

9. Which type of current is used in batteries and which is used in household outlets?

____

____

____

____

VOLTAGE

_____**10.** What is the potential difference between two points in a circuit called?
a. resistance
b. current
c. voltage
d. charge

_____**11.** Which letter is used to represent voltage in equations?
a. *G*
b. *C*
c. *V*
d. *I*

_____**12.** What happens to electric current if voltage becomes larger?
a. The current decreases.
b. The current increases.
c. The current stays the same.
d. No current will flow.

RESISTANCE

_____**13.** What is the opposition to the flow of electric charge called?
a. resistance
b. current
c. voltage
d. charge

_____**14.** Which letter is used to represent resistance in equations?
a. *E*
b. *C*
c. *V*
d. *R*

15. Resistance is expressed in ____________________.

16. The higher the resistance of a material is, the lower the

______________________.

17. What four things determine an object's resistance?

____

____

____

____

18. Good conductors have a(n) ______________________ resistance.

19. Why are high resistance materials useful in light bulbs?

____

____

20. What is a superconductor?

____

____

GENERATING ELECTRICAL ENERGY

Match the correct definition with the correct term. Write the letter in the space provided.

______ **21.** This changes chemical or radiant energy into electrical energy.

______ **22.** This is a mixture of chemicals in a cell.

______ **23.** This converts light energy into electrical energy.

______ **24.** This conducts thermal energy into electrical energy.

______ **25.** This is the part of the cell through which charges enter and exit.

a. cell
b. electrode
c. thermocouple
d. electrolyte
e. photocell

26. Chemical changes between the electrolyte and the electrodes convert

______________________ into ______________________.

27. Compare electrolytes found in a wet cell with those found in a dry cell.

____

____

28. Thermocouples are useful for monitoring the temperatures of what three things?

____

____

____

29. How do photocells convert light energy into electrical energy?

____

____

____

____

Name ______________________ Class ______________ Date ____________

Skills Worksheet

Directed Reading A

Section: Electrical Calculations

CONNECTING CURRENT, VOLTAGE, AND RESISTANCE

______ **1.** What is the ratio of voltage to current?
a. electrical power
b. electrical energy
c. resistance
d. current

______ **2.** Which of the following equations is Ohm's law?
a. $V = I \times R$
b. $I = V \times R$
c. $R = I \times V$
d. $R = V \times I$

3. How did Georg Ohm study the resistances of different materials?

____

____

____

ELECTRIC POWER

______ **4.** Which of the following is the rate at which electrical energy is changed into other forms of energy?
a. electric current
b. electric power
c. voltage
d. kilowatt

______ **5.** In the formula $P = V \times I$, what does the P stand for?
a. performance
b. power
c. price
d. penny

6. Name two common units of power.

____

____

7. What happens to a light bulb as power increases?

____

____

____

8. One kilowatt is equal to ____________________ W.

MEASURING ELECTRICAL ENERGY

9. What two factors does the electric company use to determine how much a business will pay for electrical energy?

____

____

10. What is the formula for finding electrical energy?

____

11. What unit is usually used to express electrical energy?

____

12. Electric companies use a(n) ____________________ to determine how many kilowatt-hours of energy are used by a household.

13. Name two ways you can help to save energy.

____

____

Name ______________________________ Class ______________ Date ____________

Skills Worksheet

Directed Reading A

Section: Electric Circuits

1. A closed pathway where the start and end points are the same is called

a(n) ______________________.

PARTS OF AN ELECTRIC CIRCUIT

_____ **2.** Which of the following is a complete, closed path through which electric charges flow?
- **a.** electric current
- **b.** electric circuit
- **c.** energy source
- **d.** load

_____ **3.** Which of the following is NOT a basic part of a circuit?
- **a.** wires
- **b.** force
- **c.** energy source
- **d.** load

_____ **4.** Which of the following is connected to the energy source by wires and changes electrical energy into other forms of energy?
- **a.** wires
- **b.** force
- **c.** energy source
- **d.** load

5. Name four possible energy sources for a circuit.

____

____

____

____

6. A circuit is opened and closed using a(n) ______________________.

7. What will happen to the charges in a circuit when a switch is closed?

____

____

____

8. What will happen to the charges in a circuit when a switch is open?

____

____

____

TYPES OF CIRCUITS

9. All of the electrical devices in your home are ____________________ in a large circuit.

10. A circuit in which all parts are connected in a single loop is called a(n) ____________________.

11. What happens to a series circuit if you add more loads?

____

____

____

12. In a series circuit, what happens if there is a break in the circuit?

____

____

13. Why aren't series circuits a convenient way to wire a home?

____

____

____

14. A circuit in which loads are connected side by side is called a(n) ____________________.

15. Each load in a parallel circuit uses the same ____________________.

16. What happens in a parallel circuit if one load is broken or missing?

____

____

____

17. Why can you use one light or appliance at a time in a parallel circuit even if a load fails?

____

____

____

____

HOUSEHOLD CIRCUIT SAFETY

______**18.** What is the standard voltage per branch in a home in the United States?

a. 100 V
b. 110 V
c. 120 V
d. 130 V

19. Name two things that can cause a short circuit in your home.

____

____

20. What happens to a fuse to stop the flow of charges through it?

____

____

____

21. A switch that automatically opens if the current is too high is

a(n)________________________.

22. How does a ground fault circuit interrupter act like a small circuit breaker?

____

____

____

23. Name two safety measures to follow when using electrical energy.

____

____

Name ______________________ Class ______________ Date ____________

Skills Worksheet

Vocabulary and Section Summary

Electric Charge and Static Electricity

VOCABULARY

In your own words, write a definition of the following terms in the space provided.

1. law of electric charges

____

____

2. electric force

____

____

3. electric field

____

____

4. electrical conductor

____

____

5. electrical insulator

____

____

6. static electricity

____

____

7. electric discharge

____

____

SECTION SUMMARY

Read the following section summary.

- The law of electric charges states that like charges repel and opposite charges attract.
- The size of the electric force between two objects depends on the size of the charges exerting the force and the distance between the objects.
- Charged objects exert a force on each other and can cause each other to move.
- Objects become charged when they gain or lose electrons.
- Objects may become charged by friction, conduction, or induction.
- Charges are not created or destroyed and are said to be conserved.
- Charges move easily in conductors but do not move easily in insulators.
- Static electricity is the buildup of electric charges on an object. It is lost through electric discharge.

Name ______________________ Class ____________ Date ____________

Skills Worksheet

Vocabulary and Section Summary

Electric Current and Electrical Energy

VOCABULARY

In your own words, write a definition of the following terms in the space provided.

1. electric current

____

____

2. voltage

____

____

3. resistance

____

____

4. cell

____

____

5. thermocouple

____

____

6. photocell

____

____

SECTION SUMMARY

Read the following section summary.

- Electric current is the rate at which charges pass a given point.
- An electric current can be made when there is a potential difference between two points.
- As voltage, or potential difference increases, current increases.
- An object's resistance varies depending on the object's material, thickness, length, and temperature. As resistance increases, current decreases.
- Cells and batteries convert chemical energy or radiant energy into electrical energy.
- Thermocouples and photocells are devices used to generate electrical energy.

Name ______________________ Class ______________ Date ____________

Skills Worksheet

Vocabulary and Section Summary

Electrical Calculations

VOCABULARY

In your own words, write a definition of the following term in the space provided.

1. electric power

____

____

SECTION SUMMARY

Read the following section summary.

- Ohm's law describes the relationship between current, resistance, and voltage.
- Electric power is the rate at which electrical energy is changed into other forms of energy.
- Electrical energy is electric power multiplied by time. It is usually expressed in kilowatt-hours.

Name ______________________ Class ______________ Date ______________

Skills Worksheet

Vocabulary and Section Summary

Electric Circuits

VOCABULARY

In your own words, write a definition of the following terms in the space provided.

1. series circuit

____

____

2. parallel circuit

____

____

SECTION SUMMARY

Read the following section summary.

- Circuits consist of an energy source, a load, wires, and, in some cases, a switch.
- All parts of a series circuit are connected in a single loop. The loads in a parallel circuit are on separate branches.
- Circuits fail through a short circuit or an overload. Fuses or circuit breakers protect against circuit failure.
- It is important to follow safety tips when using electrical energy.

Name ______________________ Class ______________ Date ______________

Skills Worksheet

Directed Reading A

Section: Magnets and Magnetism

PROPERTIES OF MAGNETS

1. Any material that attracts iron is a(n) ______________________.

2. The points on a magnet that have opposite magnetic qualities are the

______________________.

3. The magnetic pole that points to the north is the magnet's

______________________.

4. The magnetic pole that points to the south is the magnet's

______________________.

5. The force that can either push magnets apart or pull them together is

______________________.

6. The region around a magnet in which magnetic forces act is the

______________________.

For each description below, identify the correct magnetic property. Write either *magnetic forces* or *magnetic fields* in the space provided.

______________________ **7.** come from spinning electric charges in the magnets

______________________ **8.** can push magnets apart or pull them together

______________________ **9.** depend on how two magnets' poles line up

______________________ **10.** are regions around magnets in which magnetic forces can act

______________________ **11.** shape that can be shown with lines that surround magnets

______________________ **12.** are strongest at magnetic poles, where lines are closest together

THE CAUSE OF MAGNETISM

_____ **13.** Whether a material is magnetic depends on its
- **a.** density.
- **b.** atoms.
- **c.** shape.
- **d.** mass.

_____ **14.** As an electron moves, it makes, or induces a(n)
- **a.** aurora.
- **b.** ferromagnet.
- **c.** electromagnet.
- **d.** magnetic field.

_____ **15.** Materials in which the atoms' magnetic fields cancel each other out are
- **a.** aligned in domains.
- **b.** like iron, nickel, and cobalt.
- **c.** not magnetic.
- **d.** magnetic.

_____ **16.** Which of these is true when the poles of atoms line up?
- **a.** The atoms cancel each other out.
- **b.** The atoms are arranged in a domain.
- **c.** The atoms make a weak magnetic field.
- **d.** The atoms do not become magnetic.

17. Name one thing that causes domains of a magnet's atoms to lose alignment.

____

____

18. How do you magnetize something made of iron, cobalt, or nickel?

____

____

____

____

19. Why do you end up with two magnets when you cut one magnet in half?

____

____

____

____

KINDS OF MAGNETS

Match the correct description with the correct term. Write the letter in the space provided.

______**20.** magnet with strong magnetic properties

______**21.** magnet made by an electric current

______**22.** magnet that loses magnetization easily

______**23.** hard to magnetize, but stays magnetized

a. temporary

b. electromagnet

c. ferromagnet

d. permanent

EARTH AS A MAGNET

24. Why can magnets point north?

____

____

____

____

____

____

____

25. If you put a compass on a bar magnet, the needle points to the south pole of the magnet. Explain why.

____

____

____

____

____

____

26. Why does a compass needle point to Earth's geographic north?

____

____

____

____

____

____

27. What makes Earth's magnetic field?

____

____

____

____

____

____

28. When charged particles from the sun hit oxygen and nitrogen atoms in the air, a(n) ______________________ is formed.

Name ______________________ Class ______________ Date ____________

Skills Worksheet

Directed Reading A

Section: Magnetism from Electricity

______ **1.** What kind of train uses an electromagnet to float above the track?
a. magnetic
b. maglev
c. electric
d. electronic

THE DISCOVERY OF ELECTROMAGNETISM

______ **2.** The interaction between electricity and magnetism is called
a. electromagnetism.
b. maglev.
c. electric.
d. electronic.

______ **3.** Oersted discovered that electric current produces a(n)
a. electric field.
b. magnetic field.
c. magnetic current.
d. rotating field.

______ **4.** The direction of a magnetic field produced by an electric current depends on the direction of the
a. current.
b. magnetism.
c. wire.
d. batteries.

5. Who were the two scientists who did the first research into the interaction between electricity and magnetism?

____

____

USING ELECTROMAGNETISM

______ **6.** What are two devices that strengthen the magnetic field of a current-carrying wire?
a. magnetic field and magnetic force
b. solenoid and electromagnet
c. electromagnet and current
d. solenoid and current

______ **7.** A coil of wire that produces a magnetic field when carrying an electric current is called a(n)
a. electromagnet.
b. maglev.
c. solenoid.
d. magnetic field.

______ **8.** What happens to the magnetic field if more loops per meter are added to a solenoid?
a. The magnetic field becomes weaker.
b. The magnetic field becomes stronger.
c. The magnetic field turns on and off.
d. There is no change in the magnetic field.

______ **9.** A solenoid wrapped around a soft iron core is called a(n)
a. electromagnet.
b. maglev.
c. magnetic core.
d. magnetic field.

______ **10.** What happens to an electromagnet if the electric current in the solenoid wire is increased?
a. The electromagnet becomes weaker.
b. The electromagnet becomes stronger.
c. The electromagnet turns on and off.
d. There is no change in the electromagnet.

APPLICATIONS OF ELECTROMAGNETISM

______ **11.** What is one thing that uses an electromagnet?
a. bicycle
b. doorbell
c. computer
d. solenoid

______ **12.** An electric motor changes electrical energy into what kind of energy?
a. electromagnetic
b. electronic
c. mechanical
d. magnetic

13. Explain what happens to an electromagnet when there is no current in the wire.

____

____

____

____

Match the correct description with the correct term. Write the letter in the space provided. Some terms will not be used.

______ **14.** a device that converts electrical energy into mechanical energy

______ **15.** attached to the armature; reverses direction of electric current

______ **16.** a loop or coil of wire that can rotate

______ **17.** used to measure current

a. galvanometer

b. armature

c. electric motor

d. commutator

e. voltmeter

Name ______________________ Class ______________ Date ____________

Skills Worksheet

Directed Reading A

Section: Electricity from Magnetism

ELECTRIC CURRENT FROM A CHANGING MAGNETIC FIELD

1. What problem did both Joseph Henry and Michael Faraday work to solve?

____

____

2. The process of creating a current in a circuit by changing a magnetic field is called ______________________.

3. Describe what happened to the electric current in Michael Faraday's experiment when the battery was fully connected.

____

____

____

____

____

____

____

____

4. Describe two ways to induce a larger electric current when you move a magnet in a coil of wires.

____

____

____

____

____

____

____

____

ELECTRIC GENERATORS

_____ **5.** What device converts mechanical energy into electrical energy?
a. electric motor
b. electric generator
c. electromagnetic motor
d. magnetic motor

_____ **6.** When electric current changes direction it is called a(n)
a. generated current.
b. electromagnetic current.
c. alternating current.
d. rotating current.

7. Name the four parts of a simple generator, and describe what they do.

____

____

____

____

8. Other than the size, what is one difference between power plants and electric generators?

____

____

____

____

____

9. Name two sources of energy that generators convert into electrical energy.

____

____

Put the following steps for generating electrical energy in order from 1 to 4. Write the appropriate numbers in the space provided.

_____ **10.** Steam turns a turbine.

_____ **11.** Energy boils water into steam.

_____ **12.** Electric current is induced and electrical energy is generated.

_____ **13.** A turbine turns the magnet of a generator.

TRANSFORMERS

______ **14.** A device that increases or decreases the voltage of alternating current is called a(n)

a. voltmeter
b. generator.
c. transformer.
d. electromagnet.

15. Explain why a transformer uses different numbers of loops in its primary and secondary coils.

____

____

____

____

____

16. Describe what a step-up transformer does.

____

____

____

____

____

17. Describe what a step-down transformer does.

____

____

____

____

____

____

Name ______________________ Class ______________ Date ______________

Skills Worksheet

Vocabulary and Section Summary

Magnets and Magnetism

VOCABULARY

In your own words, write a definition of the following terms in the space provided.

1. magnet

____

____

2. magnetic pole

____

____

3. magnetic force

____

____

SECTION SUMMARY

Read the following section summary.

- All magnets have two poles. The north pole will always point to the north if allowed to rotate freely. The other pole is called the south pole.
- Like magnetic poles repel each other. Opposite magnetic poles attract.
- Every magnet is surrounded by a magnetic field. The shape of the field can be shown with magnetic field lines.
- A material is magnetic if its domains line up.
- Magnets can be classified as ferromagnets, electromagnets, temporary magnets, and permanent magnets.
- Earth acts as if it has a big bar magnet through its core. Compass needles and the north poles of magnets point to Earth's magnetic south pole, which is near Earth's geographic North Pole.
- Auroras are most commonly seen near Earth's magnetic poles because Earth's magnetic field bends inward at the poles.

Name ______________________ Class ______________ Date ____________

Skills Worksheet

Vocabulary and Section Summary

Magnetism from Electricity

VOCABULARY

In your own words, write a definition of the following terms in the space provided.

1. electromagnetism

____

____

2. solenoid

____

____

3. electromagnet

____

____

4. electric motor

____

____

SECTION SUMMARY

Read the following section summary.

- Oersted discovered that a wire carrying a current makes a magnetic field.
- Electromagnetism is the interaction between electricity and magnetism.
- An electromagnet is a solenoid that has an iron core.
- A magnet can exert a force on a wire carrying a current.
- A doorbell, an electric motor, and a galvanometer all make use of electromagnetism.

Name ______________________ Class ______________ Date ____________

Skills Worksheet

Vocabulary and Section Summary

Electricity from Magnetism

VOCABULARY

In your own words, write a definition of the following terms in the space provided.

1. electromagnetic induction

____

____

2. electric generator

____

____

3. transformer

____

____

SECTION SUMMARY

Read the following section summary.

- Electromagnetic induction is the process of making an electric current by changing a magnetic field.
- An electric generator converts mechanical energy into electrical energy through electromagnetic induction.
- A step-up transformer increases the voltage of an alternating current. A step-down transformer decreases the voltage.
- The side of a transformer that has the greater number of loops has the higher voltage.

Name ______________________ Class ______________ Date ____________

Skills Worksheet

Directed Reading A

Section: Electronic Devices

INSIDE AN ELECTRONIC DEVICE

______ 1. What role does a circuit board play?
- a. receives information from a TV
- b. connects the parts of a circuit
- c. acts like an antenna
- d. deflects current away from the circuit

______ 2. The LED in a remote control
- a. gives off radio waves.
- b. is only for decoration.
- c. sends information to the TV.
- d. receives information from the TV.

3. An LED, or ______________________, is one of the electronic components within a TV remote control.

4. A sheet of insulating material called a(n) ______________________ carries circuit elements and is inserted into electronic devices.

SEMICONDUCTORS

5. An element or compound that conducts an electric current better than an insulator does is called what?

____

____

6. What happens when silicon atoms bond?

____

____

7. The addition of an impurity element to a semiconductor is called

______________________.

8. What happens when a silicon atom is replaced with an arsenic atom? What type of semiconductor is produced?

____

____

9. What happens when a silicon atom is replaced with a gallium atom? What kind of semiconductor is produced?

____

____

DIODES

10. An electronic component that allows electric charge to move mainly in one direction is called a(n) ____________________.

11. What happens to the "extra" electrons when two layers of a diode meet?

____

____

12. Power plants send electrical energy to homes by means of what?

____

____

13. How do diodes help change AC to DC?

____

____

____

____

TRANSISTORS

______ **14.** A transistor is an electronic component that

- **a.** decreases current.
- **b.** does not affect current.
- **c.** amplifies or increases current.
- **d.** blocks current.

15. A transistor can be used in many devices, including amplifiers, oscillators, and ____________________.

16. Why is a transistor useful in an amplifier?

____

____

17. How are transistors used as switches in devices ?

____

____

INTEGRATED CIRCUITS

______ **18.** Which of the following statements about integrated circuits is true?

a. Few circuits can fit onto one integrated circuit.
b. Devices that use integrated circuits can run at very high speeds.
c. An integrated circuit has many components on several semiconductors.
d. Electric charges moving through integrated circuits have to travel great distances.

19. An entire circuit with many components on a single semiconductor is called

a(n) ______________________.

20. What is one advantage of replacing vacuum tubes with transistors and diodes?

____

____

Name ______________________ Class ______________ Date ____________

Skills Worksheet

Directed Reading A

Section: Communication Technology

COMMUNICATING WITH SIGNALS

______ **1.** One of the first electronic communication devices was the

a. telephone.
b. computer.
c. telegraph.
d. typewriter.

2. Anything that can be used to send information is called a(n)

______________________.

3. Sometimes, a signal is sent using another signal called a(n)

______________________.

ANALOG SIGNALS

4. A signal whose properties change without a break between values is called

a(n) ______________________.

5. How can the analog signal in a telephone system be described?

____

6. In a telephone, the ______________________ changes the analog signal back into the sound of your voice.

7. The ______________________ of a telephone changes sound waves into an analog signal when you speak

8. In vinyl records, the number and depth of the ______________________ in the disk represent the sound's ______________________.

9. When you play a record, the ______________________, or ______________________, makes the electromagnet vibrate.

10. Over time, what happens as a result of a stylus repeatedly touching a record?

____

____

DIGITAL SIGNALS

Match the correct description with the correct term. Write the letter in the space provided.

______**11.** represented by a pulse

______**12.** represented by a missing pulse

______**13.** represented as a sequence of separate values

______**14.** means "two"

______**15.** is short for binary digit

a. digital signal
b. binary
c. number 0
d. number 1
e. bit

16. What do the pits and lands on a compact disc do?

____

____

17. In a digital recording, the original sound is represented

by ______________________.

18. How does a CD player work?

____

____

____

RADIO AND TELEVISION

______**19.** In radio, a modulator
a. changes sound waves into electric current.
b. strengthens the analog signal.
c. combines the amplified analog signal with radio waves.
d. removes the radio waves from the analog signal.

______**20.** Which of the following transmits modulated radio waves through the air?
a. microphone
b. radio tower
c. antenna
d. modulator

21. TV and radio signals can be either ______________________ or analog.

22. Define *electromagnetic wave.*

____

____

23. What role do electromagnetic waves play in radio and television broadcasts?

____

24. Television images are made by beams of ______________ hitting the screen.

25. What are three ways in which signals are sent to your TV?

____

____

____

26. Why is it better to watch digital shows on a digital display rather than on an analog display?

____

____

____

27. Video signals transmitted from a TV station are received by the ______________ of a TV receiver.

28. What is the role of fluorescent materials in producing an image on a color television?

____

____

29. Why are standard television sets so bulky and heavy?

____

____

____

30. New types of television screens, called ______________ ______________, are thinner than standard screens.

31. In a plasma display, a(n) ______________ charges thousands of cells with gases in them to produce a current in the gases.

32. In a plasma display, each well contains fluorescent materials that give off what?

____

Name ______________________ Class ______________ Date ____________

Skills Worksheet

Directed Reading A

Section: Computers

WHAT IS A COMPUTER?

Match the correct definition with the correct term. Write the letter in the space provided.

_____ **1.** performing an action, such as adding a list of numbers

_____ **2.** information you give a computer

_____ **3.** final result of the work done by the computer

_____ **4.** information in a computer's memory

_____ **5.** electronic device that performs tasks by following instructions

a. output
b. processing
c. storage
d. input
e. computer

6. What was the first general-purpose computer called?

____

7. Who made the first general-purpose computer, and when was it built?

____

____

8. Why did ENIAC have to be cooled?

____

____

____

9. A single semiconductor chip called a(n) ______________________ controls and carries out a computer's instructions.

10. What component has made it possible for modern computers to become so small?

____

____

____

COMPUTER HARDWARE

______ **11.** The parts or pieces of equipment that make up a computer are called
a. hardware.
b. software.
c. processes.
d. hardwires.

Match the correct description with the correct term. Write the letter in the space provided.

______ **12.** permanent memory

______ **13.** device used to send information through telephone lines

______ **14.** a type of output device

______ **15.** temporary memory

______ **16.** a type of input device

______ **17.** the microprocessor in a personal computer

a. mouse
b. CPU
c. ROM
d. RAM
e. printer
f. modem

COMPACT DISCS

18. What can be stored on a CD?

____

____

____

Match the correct description with the correct term. Write the letter in the space provided.

______ **19.** device used to "burn" or heat dyes on recordable compact discs

______ **20.** computer disc that can be used only once

______ **21.** computer disc that can be erased and written over again

a. CD-R
b. CD-RW
c. laser

COMPUTER SOFTWARE

22. What is a set of instructions or commands for a computer called?

____

23. What are three jobs that are handled by operating-system software?

____

____

____

24. What does application software tell the computer to do?

____

____

COMPUTER NETWORKS

Match the correct description with the correct term. Write the letter in the space provided.

______**25.** network in which groups of computers connect to an ISP through only one line

______**26.** huge computer network made up of millions of computers

______**27.** collection of pages that share a simple format

______**28.** feature used to find Web pages

______**29.** area on a Web page that is clicked on to move from one page or site to another

______**30.** program used to view pages on the Internet

a. link

b. Web browser

c. LAN

d. search engine

e. Internet

f. World Wide Web

Name ______________________ Class ______________ Date ____________

Skills Worksheet

Vocabulary and Section Summary

Electronic Devices

VOCABULARY

In your own words, write a definition of the following terms in the space provided.

1. circuit board

____

____

2. semiconductor

____

____

3. doping

____

____

4. diode

____

____

5. transistor

____

____

6. integrated circuit

____

____

SECTION SUMMARY

Read the following section summary.

- Circuit boards contain circuits that supply current to different parts of electronic devices.
- Semiconductors are often used in electronic devices because their conductivity can be changed by doping.
- Diodes allow current in one direction and can change AC to DC.
- Transistors are used in amplifiers and switches.
- Integrated circuits have made smaller, smarter electronic devices possible.

Name ______________________ Class ______________ Date ____________

Skills Worksheet

Vocabulary and Section Summary

Communication Technology

VOCABULARY

In your own words, write a definition of the following terms in the space provided.

1. analog signal

____

____

2. digital signal

____

____

SECTION SUMMARY

Read the following section summary.

- Signals transmit information in electronic devices. Signals can be transmitted using a carrier. Signals can be analog or digital.
- Analog signals have continuous values. Telephones, record players, radios, and regular TV sets use analog signals.
- In a telephone, a transmitter changes sound waves to electric current. The current is sent across a phone line. The receiving telephone converts the signal back into a sound wave.
- Analog signals of sounds are used to make vinyl records. Changes in the groove reflect changes in the sound.
- Digital signals have discrete values, such as 0 and 1. CD players use digital signals.
- Radios and television sets use electromagnetic waves. These waves travel through the atmosphere. In a radio, the signals are converted to sound waves. In a television set, electron beams convert the signals into images on the screen.

Name ______________________ Class ______________ Date ____________

Skills Worksheet

Vocabulary and Section Summary

Computers

VOCABULARY

In your own words, write a definition of the following terms in the space provided.

1. computer

____

____

2. microprocessor

____

____

3. hardware

____

____

4. software

____

____

5. Internet

____

____

SECTION SUMMARY

Read the following section summary.

- All computers have four basic functions: input, processing, storage, and output.
- The first general-purpose computer, ENIAC, was made of thousands of vacuum tubes and filled an entire room. Microprocessors have made it possible to have computers the size of notebooks.
- Computer hardware includes input devices, the CPU, memory, output devices, and modems.
- CD burners can store information on recordable CDs, or CD-Rs. Rewritable CDs, or CD-RWs, can be erased and reused. Both use patterns of light and dark spots.
- Computer software is a set of instructions that tell a computer what to do. The two main types are operating systems and applications. Applications include word processors, spreadsheets, and games.
- The Internet is a huge network that allows millions of computers to share information.

Name ______________________ Class ______________ Date ____________

Skills Worksheet

Directed Reading A

Section: The Nature of Waves

1. What is a wave?

____

____

____

WAVE ENERGY

2. Why does a wave move toward the shore but the leaf floating on the surface of the water does not?

____

____

3. A substance through which a wave can travel is called a(n)

______________________.

4. How is energy transmitted through a medium?

____

____

____

____

5. Define *vibration.*

____

____

6. Name three types of waves that require a medium.

____

____

____

7. Waves that require a medium are called ______________________.

8. List three examples of waves that can transfer energy without going through a medium.

____

____

____

9. Waves that do not require a medium but can go through matter are called

______________________.

TYPES OF WAVES

10. What are the two main types of waves?

____

____

11. How does a surface wave form?

____

____

Match the correct description with the correct term. Write the letter in the space provided.

______**12.** a part of a longitudinal wave where the particles are crowded together

______**13.** word that means to be "at right angles"

______**14.** the highest point of a transverse wave

______**15.** a part of a longitudinal wave where the particles are spread apart

______**16.** a wave in which the particles of the medium move perpendicularly to the direction the wave is traveling

______**17.** the lowest point between each crest of a transverse wave

______**18.** a wave in which the particles of the medium move back and forth along the path that the wave moves

a. transverse wave
b. compression
c. trough
d. perpendicular
e. rarefaction
f. crest
g. longitudinal wave

Name ______________________________ Class ______________ Date ____________

Skills Worksheet

Directed Reading A

Section: Properties of Waves

______ **1.** The height of waves and the distance between crests are examples of
a. wave interactions.
b. media.
c. wave properties.
d. rest positions.

AMPLITUDE

______ **2.** The amplitude of a wave is related to its
a. speed.
b. frequency.
c. length.
d. height.

______ **3.** The point where particles in a medium stay when there are no disturbances is the
a. rest position.
b. wavelength.
c. amplitude.
d. compression.

4. Define *amplitude.*

____

____

____

5. A wave with a(n) ______________________ amplitude carries more energy than a wave with a(n) ______________________ amplitude.

WAVELENGTH

______ **6.** The distance from any point on a wave to an identical point on the next wave is called a
a. crest.
b. wavelength.
c. trough.
d. frequency.

7. A wave with a(n) ______________________ wavelength carries more energy than a wave with a(n) ______________________ wavelength.

FREQUENCY

______ **8.** The number of waves produced in a given amount of time is known as the wave's
a. hertz. **c.** frequency.
b. speed. **d.** compression.

______ **9.** Which of the following units are used to measure frequency?
a. meters **c.** decibels
b. hertz **d.** watts

______ **10.** One hertz equals how many waves per second?
a. 1 **c.** 10
b. 5 **d.** 100

11. If the amplitudes of waves are equal, ______________________ waves carry more energy than ______________________ waves.

WAVE SPEED

______ **12.** What is the speed at which a wave travels through a medium called?
a. frequency **c.** wavelength
b. hertz **d.** wave speed

______ **13.** If the speed and the wavelength of a particular wave is known, then the wave equation can be used to determine the wave's
a. amplitude. **c.** compression.
b. frequency. **d.** hertz.

______ **14.** If a wave is traveling at a certain speed and its frequency is cut in half, what happens to the wavelength?
a. The wavelength is doubled.
b. The wavelength is halved.
c. The wavelength remains the same.
d. The wavelength is inverted.

15. Wave speed is equal to ______________________ multiplied by frequency.

16. How is the wave equation used to calculate wave speed written?

____

Name ______________________ Class ______________ Date ______________

Skills Worksheet

Directed Reading A

Section: Wave Interactions

REFLECTION

1. Why do planets and the moons shine so brightly if they do not produce light?

____

____

2. When a wave bounces back after hitting a barrier, ______________________ occurs.

3. A reflected sound wave is called a(n) ______________________.

4. When a wave passes through a substance, the wave is ______________________ through that substance.

REFRACTION

5. When a wave bends as it passes from one medium to another at an angle, ______________________ occurs.

6. What causes a wave to bend and travel in new direction when it moves from one medium to another?

____

____

____

____

7. Why can you see a rainbow when sunlight is refracted through water droplets?

____

____

DIFFRACTION

8. Define *diffraction.*

____

____

____

9. If the barrier or opening is larger than the wavelength of the wave, there is only a(n) ______________________ amount of diffraction.

10. Why can you hear around corners but cannot see around corners?

____

____

____

INTERFERENCE

11. What is the combination of two or more waves that results in a single wave called?

____

12. What is it called when a combined wave has a larger amplitude than the original waves?

____

13. What is it called when a combined wave has a smaller amplitude than the original waves?

____

14. In a(n) ______________________, certain parts of the wave are always at the rest position because of total destructive interference between all the waves.

15. The frequencies at which standing waves are made are called

______________________.

16. When an object vibrating at or near the resonant frequency of a second object causes the second object to vibrate, ______________________ occurs.

17. How does resonance work when you sing in the shower?

____

____

____

Name ______________________ Class ______________ Date ____________

Skills Worksheet

Vocabulary and Section Summary

The Nature of Waves

VOCABULARY

In your own words, write a definition of the following terms in the space provided.

1. wave

____

____

2. medium

____

____

3. transverse wave

____

____

4. longitudinal wave

____

____

SECTION SUMMARY

Read the following section summary.

- A wave is a disturbance that transmits energy.
- The particles of a medium do not travel with the wave.
- Mechanical waves require a medium, but electromagnetic waves do not.
- Particles in a transverse wave vibrate perpendicularly to the direction the wave travels.
- Particles in a longitudinal wave vibrate parallel to the direction that the wave travels.

Name ______________________ Class ______________ Date ____________

Skills Worksheet

Vocabulary and Section Summary

Properties of Waves

VOCABULARY

In your own words, write a definition of the following terms in the space provided.

1. amplitude

____

____

2. wavelength

____

____

3. frequency

____

____

4. wave speed

____

____

SECTION SUMMARY

Read the following section summary.

- Amplitude is the maximum distance the particles of a medium vibrate from their rest position.
- Wavelength is the distance between two adjacent corresponding parts of a wave.
- Frequency is the number of waves that pass a given point in a given amount of time.
- Wave speed can be calculated by multiplying the wave's wavelength by the frequency.

Name ______________________ Class ______________ Date ____________

Skills Worksheet

Vocabulary and Section Summary

Wave Interactions

VOCABULARY

In your own words, write a definition of the following terms in the space provided.

1. reflection

____

____

2. refraction

____

____

3. diffraction

____

____

4. interference

____

____

5. standing wave

____

____

6. resonance

____

____

SECTION SUMMARY

Read the following section summary.

- Waves reflect after hitting a barrier.
- Refraction is the bending of a wave when it passes through different media.
- Waves bend around barriers or through openings during diffraction.
- The result of two or more waves overlapping is called interference.
- Amplitude increases during constructive interference and decreases during destructive interference.
- Resonance occurs when a vibrating object causes another object to vibrate at one of its resonant frequencies.

Name ______________________ Class ______________ Date ______________

Skills Worksheet

Directed Reading A

Section: What Is Sound?

SOUND AND VIBRATIONS

1. The complete back-and-forth motion of an object is called a(n)

vibrations.

2. In a(n) compressions the particles in the air are closer together than in the surrounding air.

3. In a(n) rarefactors, the particles in the air are less crowded than in the surrounding air.

4. Longitudinal waves that are caused by vibrations and that travel through a medium are called sound wave.

5. As sound waves leave their source, in what direction do they travel?

sound waves travel in all directions away from their source

6. Air does not travel with sound waves. But what would happen at the school dance if air did travel with sound?

If the air did travel with sound, wind gusts from music speakers would blow you over at a school dance

7. A substance through which a wave can travel is a(n)

medium.

8. Why is there no sound in a vacuum?

because there are no particles to vibrate

9. Why does the sound of the ringing alarm clock get quieter as the air is removed?

Tubing is connected to a pump that is removing air from the jar.

HOW YOU DETECT SOUND

C 10. After your ears convert sound waves into electrical signals, where are the signals sent for interpretation?

a. pinna
b. spinal cord
c. brain
d. oval window

Match the labels to the parts of the drawing. Write the letters in the spaces provided.

__g__ **11.** ear canal

__a__ **12.** pinna

__e__ **13.** cochlea

__f__ **14.** eardrum

__b__ **15.** hammer

__c__ **16.** anvil

__d__ **17.** stirrup

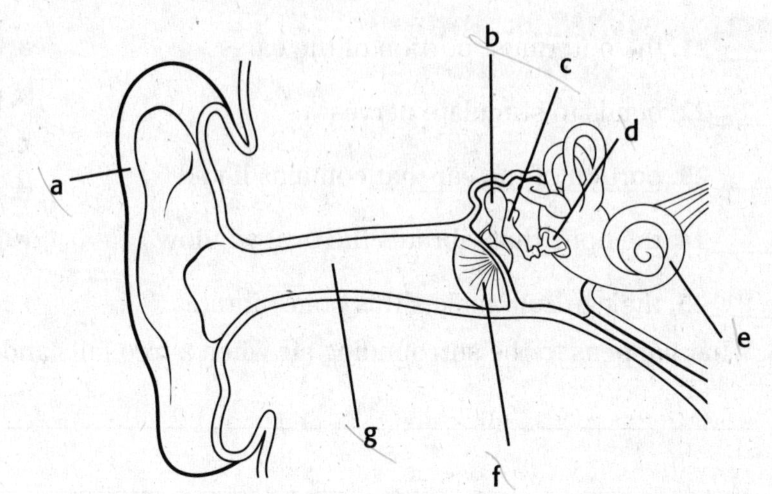

Match the labels to the parts of the drawing. Write the letters in the spaces provided.

__C__ **18.** middle ear

__A__ **19.** outer ear

__B__ **20.** inner ear

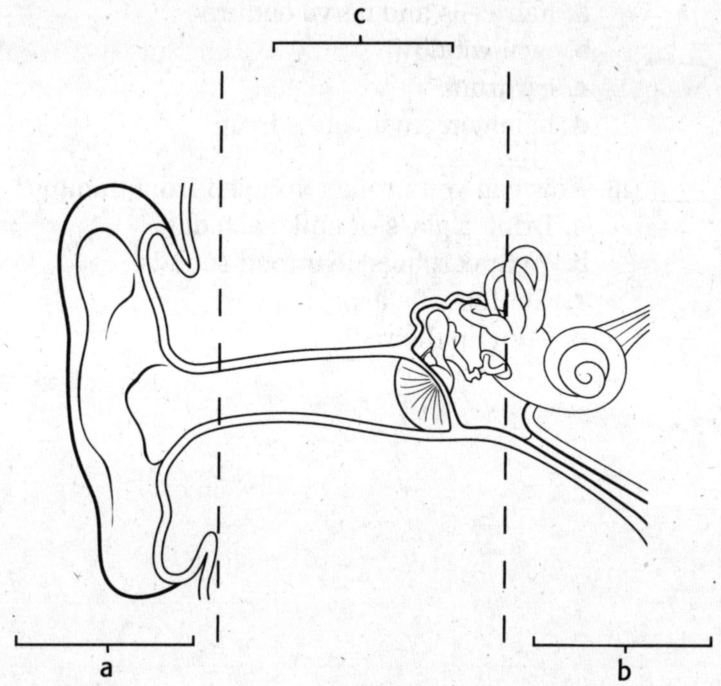

Match the correct definition with the correct term. Write the letter in the space provided.

b 21. the outermost portion of the ear

d 22. bends to stimulate nerves

e 23. portion of the ear that contains liquid

a 24. the bone that vibrates the oval window

c 25. the eardrum makes this bone vibrate

a. cochlea
b. pinna
c. hammer
d. stirrup
e. hair cell

26. What happens to the surrounding air when a tree falls and hits the ground?

these vibrations make compressions and rarefactions in the surrounding air

HEARING LOSS AND DEAFNESS

A **27.** Loud sounds can cause damage to the
a. hair cells and nerve endings.
b. oval window.
c. eardrum.
d. hammer, anvil, and stirrup.

B **28.** How can you protect yourself from tinnitus?
a. Drink a glass of milk each day.
b. Wear earplugs near loud sounds.
c. Get lots of sleep.
d. Turn up the radio.

Name ______________________ Class ______________ Date ______________

Skills Worksheet

Directed Reading A

Section: Properties of Sound

1. The differences between sounds depend on the properties of the

sound waves.

THE SPEED OF SOUND

B **2.** How quickly a sound reaches your ears depends on
- **a.** how loud or soft the sound is.
- **b.** the medium through which the sound is traveling.
- **c.** what causes the sound.
- **d.** the properties of the sound.

3. In general, what happens to the speed of sound as a medium cools?

the slower the speed of sound

4. What happens to particles as they slow down?

more slowly and transmit energy more slower than particles do in warmer material

5. What did Chuck Yeager accomplish in 1947?

the X-1 airplane was the first vehicle to move faster than the speed of sound

PITCH AND FREQUENCY

A **6.** A measure of how high or how low a sound is perceived to be is
- **a.** its pitch.
- **b.** its frequency.
- **c.** its speed.
- **d.** its medium.

D **7.** Pitch is NOT related to
- **a.** the frequency of the sound wave.
- **b.** the number of Hertz of the sound.
- **c.** the number of sound waves produced in a given time.
- **d.** how far away the source of the sound is from your ear.

D **8.** The sound produced by a dog whistle
- **a.** has a frequency too low for people to hear.
- **b.** has a pitch too low for people to hear.
- **c.** cannot be heard by a dog.
- **d.** is called an ultrasonic sound.

9. The apparent change in the frequency of a sound caused by the motion of either the listener or the source of the sound is the Doppler effect.

10. What happens to the sound waves from a moving source, such as a car with its horn honking, when the sound waves are moving in the same direction as the car?

the compressions and rarefactors of the sound wave will be closer together

11. How do the frequency and pitch of the sound seem to a person in front of a moving car with its horn honking?

seem to be high; is moving in the opposite direction that the sound waves are moving

12. How do the frequency and pitch of the sound seem to a person behind a moving car with its horn honking?

Seems to be low,

13. What happens to the pitch of the sound that the driver hears?

the driver always hears the same pitch because the driver is moving with the car.

LOUDNESS AND AMPLITUDE

Match the correct description with the correct term. Write the letter in the space provided.

d **14.** the unit used to express how loud or soft a sound is perceived

b **15.** how loud or soft a sound is perceived

c **16.** the maximum distance the particles in a wave vibrate from their rest positions

a. decibel
b. loudness
c. amplitude

Name ______________________ Class ______________ Date ______________

"SEEING" AMPLITUDE AND FREQUENCY

17. What does an oscilloscope do?

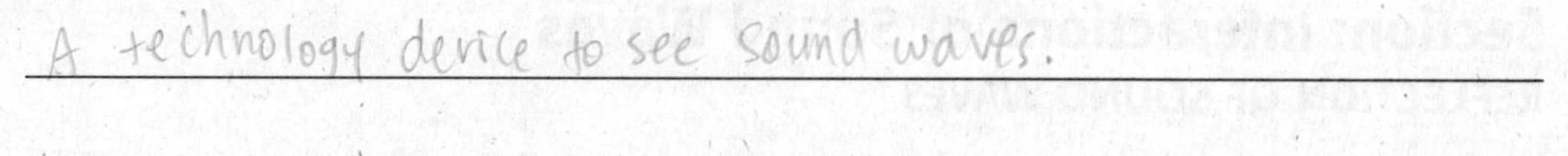

____

____

____

Look at the two sounds represented on the oscilloscope screens below. Then, answer the following questions.

low pitch
lower frequency

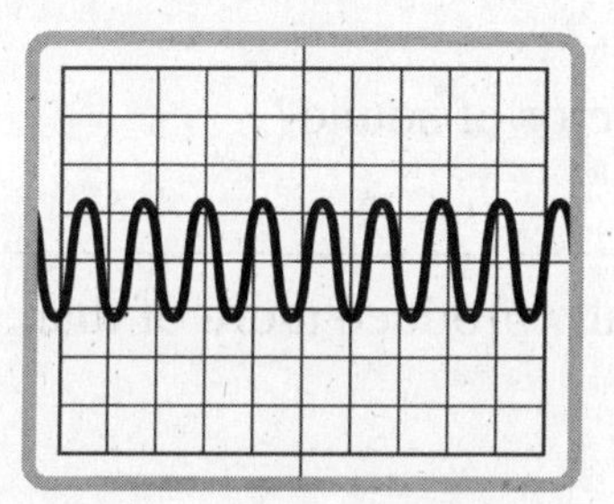

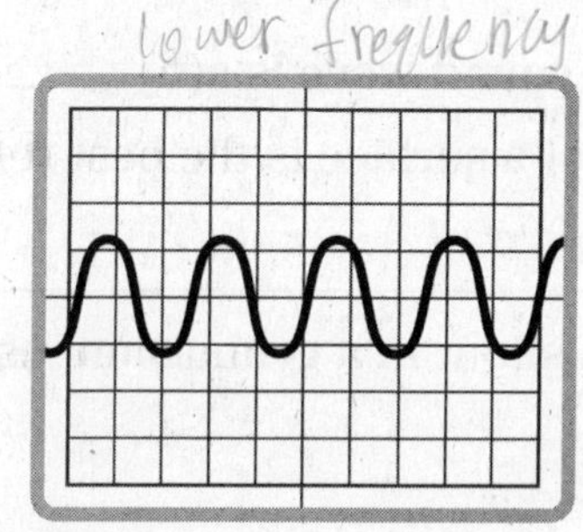

18. How is the frequency of the sounds different?

____

____

19. How is the pitch of the sounds different?

____

____

20. What does the line on a graph from an oscilloscope represent?

the sound ways

____

Name ______________________ Class ____________ Date ____________

Skills Worksheet

Directed Reading A

Section: Interactions of Sound Waves

REFLECTION OF SOUND WAVES

1. List two reasons why sounds are important to beluga whales.

to communicate and find food

____

2. The bouncing back of a wave after it strikes a barrier is called a(n) Reflections.

3. A reflected sound wave is a(n) echo.

4. What kind of a surface is the best reflector of sound?

best off smooth, hard surface

5. What will a shout in a gymnasium usually produce more of than a shout in an auditorium?

In a gymnasium can produce echo

6. The use of reflected sounds by animals such as bats to find objects is called echolocation.

7. How does the Doppler effect help bats find food?

The bat can detect an insect flying toward

8. The process of using reflected sound waves to find objects is called echolocation.

9. What are three ways that sonar is used?

① ultrasonic waves

② to help navigators on ships avoid icebergs

③ oceangraphers map the ocean floor

10. The term for a medical procedure that uses echoes to "see" inside a patient's body is called Ultrasonography.

11. What are three ways that ultrasonic waves can be used in medicine?

① patient's internal organs

② to check the devlopment of an unborn baby

③ ultrasonic waves are less harmful than x-rays are

INTERFERENCE OF SOUND WAVES

12. When two or more waves combine to form a single wave,

in terference occurs.

13. In destructive interference, two sound waves combine so that the compressions of one wave overlap the rarefactions of another wave to produce a softer sound.

14. In constructive interference, two sound waves combine so that the compressions of one wave overlap the compressions of another wave to produce a louder sound.

15. When an airplane travels faster than the speed of sound, a(n)

sonic boom is created.

16. Explain what happens when a jet flies at supersonic speeds.

the sound waves it creates spread out behind it in a 3-dimensional cone shape

17. In a(n) standing wave, a pattern of vibration looks like a wave that is standing still.

RESONANCE

18. What is resonance?

____

19. Under what circumstances can a tuning fork cause a guitar string to vibrate without touching it?

Resonance

20. How does a guitar use resonance to make sound?

when the string vibrates, sound waves enter the body of guitar

Name ________________ Class ________ Date ________

Skills Worksheet

Directed Reading A

Section: Sound Quality

__D__ **1.** What is the difference between music and sound?

a. loudness
b. pitch
c. amplitude
d. sound quality

WHAT IS SOUND QUALITY?

2. Why do the same notes sound different on different instruments?

Because of different sound, and frenquency

3. The result of several pitches mixing together through interference is sound quality.

SOUND QUALITY OF INSTRUMENTS

__C__ **4.** What causes differences in sound quality among different musical instruments?

a. interference
b. frequency
c. structural differences
d. standing waves

Match the correct instrument with the correct family. Write the letter in the space provided. Each family will be used more than once.

a. string instrument
b. wind instrument
c. percussion instrument

__C__ **5.** drum
__a__ **6.** guitar
__b__ **7.** trumpet
__a__ **8.** cello
__C__ **9.** bells
__b__ **10.** tuba
__b__ **11.** clarinet
__a__ **12.** banjo
__C__ **13.** cymbals
__C__ **14.** saxophone
__a__ **15.** violin

Fill in each blank in questions 16 through 18 using either *lower pitch* or *higher pitch*.

16. In a string instrument, a thicker string has a(n) lower pitch.

17. In a wind instrument, lengthening the air column produces a(n) vibrations.

18. Among percussion instruments, larger instruments produce a(n) the lower the pitch is.

MUSIC OR NOISE?

B **19.** Which of the following would produce a sound wave with a repeating pattern?

a. slamming a door
b. French horn
c. keys falling to the floor
d. truck engine

20. A sound that consists of a random mix of frequencies is a(n) noise.

21. What is the difference between the two sound waves shown in the oscilloscopes below?

repeated notes

French horn

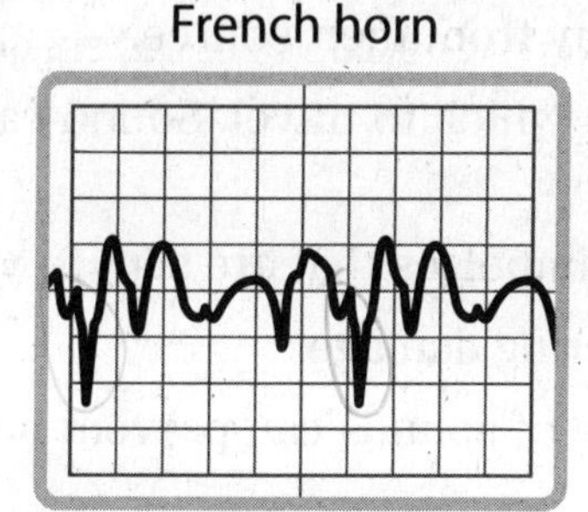

complex notes

A sharp clap

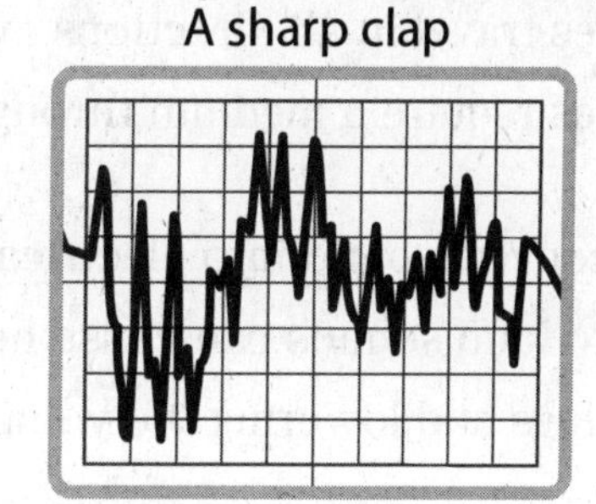

____

____

____

Name ______________________ Class ______________ Date ____________

Skills Worksheet

Vocabulary and Section Summary

What Is Sound?

VOCABULARY

In your own words, write a definition of the following terms in the space provided.

1. sound wave

____

____

2. medium

____

____

SECTION SUMMARY

Read the following section summary.

- All sounds are generated by vibrations.
- Sounds travel as longitudinal waves consisting of compressions and rarefactions.
- Sound waves travel in all directions away from their source.
- Sound waves require a medium through which to travel. Sound cannot travel in a vacuum.
- Your ears convert sound into electrical impulses that are sent to your brain.
- Exposure to loud sounds can cause hearing damage.
- Using earplugs and lowering the volume of sounds can prevent hearing damage.

Name ______________________ Class ______________ Date ____________

Skills Worksheet

Vocabulary and Section Summary

Properties of Sound

VOCABULARY

In your own words, write a definition of the following terms in the space provided.

1. pitch

____

____

2. Doppler effect

____

____

3. loudness

____

____

4. decibel

____

____

SECTION SUMMARY

Read the following section summary.

- The speed of sound depends on the medium and the temperature.
- The pitch of a sound becomes higher as the frequency of the sound wave becomes higher. Frequency is expressed in units of Hertz (Hz), which is equivalent to waves per second.
- The Doppler effect is the apparent change in frequency of a sound caused by the motion of either the listener or the source of the sound.
- Loudness increases with the amplitude of the sound. Loudness is expressed in decibels.
- The amplitude and frequency of a sound can be measured electronically by an oscilloscope.

Name ______________________ Class ______________ Date ____________

Skills Worksheet

Vocabulary and Section Summary

Interactions of Sound Waves

VOCABULARY

In your own words, write a definition of the following terms in the space provided.

1. echo

____

____

2. echolocation

____

____

3. interference

____

____

4. sonic boom

____

____

5. standing wave

____

____

6. resonance

____

____

SECTION SUMMARY

Read the following section summary.

- Echoes are reflected sound waves.
- Some animals can use echolocation to find food or to navigate around objects.
- People use echolocation technology in many underwater applications.
- Ultrasonography uses sound reflection for medical applications.
- Sound barriers and shock waves are created by interference.
- Standing waves form at an object's resonant frequencies.
- Resonance happens when a vibrating object causes a second object to vibrate at one of its resonant frequencies.

Name ______________________ Class ______________ Date ____________

Skills Worksheet

Vocabulary and Section Summary

Sound Quality

VOCABULARY

In your own words, write a definition of the following terms in the space provided.

1. sound quality

____

____

2. noise

____

____

SECTION SUMMARY

Read the following section summary.

- Different instruments have different sound qualities.
- Sound quality results from the blending through interference of the fundamental and several overtones.
- The three families of instruments are string instruments, wind instruments, and percussion.
- Noise is a sound consisting of a random mix of frequencies.

Name ____________________ Class ____________ Date ____________

Skills Worksheet

Vocabulary and Section Summary

What Is Light?

VOCABULARY

In your own words, write a definition of the following terms in the space provided.

1. electromagnetic wave

____

____

2. radiation

____

____

SECTION SUMMARY

Read the following section summary.

- Light is an electromagnetic (EM) wave. An EM wave is a wave that consists of changing electric and magnetic fields. EM waves require no matter through which to travel.
- EM waves can be produced by the vibration of charged particles.
- The speed of light in a vacuum is about 300,000,000 m/s.
- EM waves from the sun are the major source of energy for Earth.

Name ______________________ Class ______________ Date ____________

Skills Worksheet

Vocabulary and Section Summary

The Electromagnetic Spectrum

VOCABULARY

In your own words, write a definition of the following term in the space provided.

1. electromagnetic spectrum

____

____

SECTION SUMMARY

Read the following section summary.

- All electromagnetic (EM) waves travel at the speed of light. EM waves differ only by wavelength and frequency.
- The entire range of EM waves is called the *electromagnetic spectrum.*
- Radio waves are used for communication.
- Microwaves are used in cooking and in radar.
- The absorption of infrared waves is felt as an increase in temperature.
- Visible light is the narrow range of wavelengths that humans can see. Different wavelengths are seen as different colors.
- Ultraviolet light is useful for killing bacteria and for producing vitamin D in the body. Overexposure to ultraviolet light can cause health problems.
- X rays and gamma rays are EM waves that are often used in medicine. Overexposure to these kinds of rays can damage or kill living cells.

Name ______________________ Class ______________ Date ______________

Skills Worksheet

Vocabulary and Section Summary

Interactions of Light Waves

VOCABULARY

In your own words, write a definition of the following terms in the space provided.

1. reflection

____

____

2. absorption

____

____

3. scattering

____

____

4. refraction

____

____

5. diffraction

____

____

6. interference

____

____

SECTION SUMMARY

Read the following section summary.

- The law of reflection states that the angle of incidence is equal to the angle of reflection.
- Things that are luminous can be seen because they produce their own light. Things that are illuminated can be seen because light reflects off them.
- Absorption is the transfer of light energy to particles of matter. Scattering is an interaction of light with matter that causes light to change direction.
- Refraction of light waves can create optical illusions and can separate white light into separate colors.
- How much light waves diffract depends on the light's wavelength. Light waves diffract more when traveling through a narrow opening.
- Interference can be constructive or destructive. Interference of light waves can cause bright and dark bands.

Name ______________________ Class ______________ Date ______________

Skills Worksheet

Vocabulary and Section Summary

Light and Color

VOCABULARY

In your own words, write a definition of the following terms in the space provided.

1. transmission

____

____

2. transparent

____

____

3. translucent

____

____

4. opaque

____

____

5. pigment

____

____

SECTION SUMMARY

Read the following section summary.

- Objects are transparent, translucent, or opaque depending on their ability to transmit light.
- Colors of opaque objects are determined by the color of light that they reflect.
- Colors of translucent and transparent objects are determined by the color of light they transmit.
- White light is a mixture of all colors of light.
- Light combines by color addition. The primary colors of light are red, blue, and green.
- Pigments give objects color. Pigments combine by color subtraction. The primary pigments are magenta, cyan, and yellow.

Name ______________________ Class ______________ Date ____________

Skills Worksheet

Directed Reading A

Section: Mirrors and Lenses

RAYS AND THE PATH OF LIGHT WAVES

______ **1.** Light waves travel from a light source in a
- **a.** curved line.
- **b.** straight line.
- **c.** wavy line.
- **d.** right angle.

______ **2.** Light waves that bounce off an object are
- **a.** reflected.
- **b.** refracted.
- **c.** deflected.
- **d.** waved.

______ **3.** Light waves that bend when passing from one medium to another are
- **a.** reflected.
- **b.** refracted.
- **c.** deflected.
- **d.** waved.

4. An arrow used to show the path and direction of a light wave is

a(n) ______________________.

5. In ray diagrams, what do rays show?

____

____

____

MIRRORS AND REFLECTION OF LIGHT

______ **6.** How many different shapes of mirror surfaces are there?
- **a.** one
- **b.** two
- **c.** three
- **d.** four

______ **7.** A mirror that has a flat surfaces is a
- **a.** plane mirror.
- **b.** concave mirror.
- **c.** convex mirror.
- **d.** curved mirror.

______ **8.** Why does the image in a plane mirror seem to come from behind the mirror?
a. because the image is reversed left to right
b. because your brain thinks light rays travel directly from an object to your eyes
c. because your brain knows the light reflects off the mirror's surface
d. because the image is a virtual image

______ **9.** A virtual image is an image through which
a. light is reflected.
b. light travels.
c. light does not travel.
d. light is absorbed.

______ **10.** The image formed by a plane mirror is a
a. virtual image.
b. real image.
c. virtual image or real image.
d. black-and-white image.

______ **11.** A mirror that is curved inward, like the inside of a spoon, is a
a. plane mirror.
b. concave mirror.
c. convex mirror.
d. wave mirror.

______ **12.** The image formed by a concave mirror is a
a. virtual image.
b. real image.
c. virtual image or a real image.
d. black-and-white image.

______ **13.** A real image is an image through which
a. light is reflected.
b. light travels.
c. light does not travel.
d. light is absorbed.

______ **14.** A mirror that is curved outward, like the back of a spoon, is a
a. plane mirror.
b. concave mirror.
c. convex mirror.
d. wave mirror.

______**15.** The image formed by a convex mirror is a
a. virtual image.
b. real image.
c. virtual image or real image.
d. black-and-white image.

Match the correct description with the correct term. Write the letter in the space provided.

______**16.** straight line drawn outward from the center of a mirror

______**17.** spot through which light rays entering the mirror parallel to the optical axis are reflected

______**18.** distance between the mirror's surface and the focal point

a. focal point
b. focal length
c. optical axis

19. Will an object more than 1 focal length away from a concave mirror form an image that is upside down or right side up? Explain.

____

____

____

20. Why are concave mirrors used in car headlights and flashlights?

____

____

____

21. What kind of images are formed by convex mirrors? Explain.

____

____

____

LENSES AND REFRACTION OF LIGHT

22. How are lenses and mirrors alike?

____

Name ______________________ Class ______________ Date ____________

Skills Worksheet

Directed Reading A

Section: Light and Technology

1. Name four types of technology that use light or other electromagnetic waves.

____

____

____

____

OPTICAL INSTRUMENTS

______ **2.** Devices that use mirrors and lenses to help people make observations are called
- **a.** laser instruments.
- **b.** telescopic instruments.
- **c.** optical instruments.
- **d.** light-wave instruments.

Match the correct description with the correct term. Write the letter in the space provided.

______ **3.** controls the amount of light that enters the camera

______ **4.** focuses light on film

______ **5.** coated with chemicals that react when exposed to light

______ **6.** opening that lets light into the camera

a. film
b. lens
c. shutter
d. aperture

7. What does a digital camera use to record images?

____

____

8. How does a refracting telescope work?

____

____

____

9. How does a reflecting telescope work?

____

____

____

10. How are light microscopes similar to refracting telescopes?

____

____

____

LASERS AND LASER LIGHT

______ **11.** A device that produces intense light of only one wavelength and color is a

a. telescope.
b. microscope.
c. laser.
d. digital camera.

12. Light waves that behave as one wave and move together as they travel from the source are called ____________________.

13. What is the difference between laser light and nonlaser light?

____

____

14. A particle of light released when an electron moves from a higher energy level to a lower energy level is called a(n) ____________________.

Match the correct definition with the correct term. Write the letter in the space provided.

______ **15.** increase in the brightness of light

______ **16.** energy transferred as electromagnetic waves

______ **17.** release of photons

______ **18.** when a photon strikes an atom and makes the atom emit a photon

a. stimulated emission
b. amplification
c. emission
d. radiation

19. A piece of film that produces a three-dimensional image of an object is a(n) ____________________.

OPTICAL FIBERS

_____ **20.** A thin glass wire that transmits light over long distances is called a(n)

a. laser fiber. **c.** optical fiber.
b. photon fiber. **d.** digital fiber.

21. What are three things optical fibers are used for?

____

____

____

22. What is total internal reflection?

____

____

____

____

POLARIZED LIGHT

23. Light waves that vibrate in only one plane are called

______________________.

24. How do polarized sunglasses reduce glare?

____

____

____

25. How does a polarizing filter work?

____

____

____

____

COMMUNICATION TECHNOLOGY

_____ **26.** Radio waves and microwaves are kinds of

a. electric waves. **c.** electromagnetic waves.
b. magnetic waves. **d.** visible waves.

Put the steps of a cordless telephone call in order from 1 to 4. Write the appropriate number in the space provided.

______ **27.** The telephone signal is changed to a radio wave and sent to the handset.

______ **28.** The handset changes the radio signal to sound.

______ **29.** The base of the telephone receives calls through the phone line.

______ **30.** The handset changes your voice to a radio wave that is sent back to the base.

31. What is one way cellular telephones are similar to cordless telephones?

____

32. What are two differences between cellular telephones and cordless telephones?

____

____

33. What is used by satellite television to transmit data?

____

34. What do satellite television companies use to broadcast microwave signals?

____

35. Why do television companies broadcast signals from space rather than from an antenna on Earth?

____

____

36. How does the Global Positioning System (GPS) work?

____

____

____

____

37. What are two uses for the Global Positioning System?

____

____

Name ______________________ Class ______________ Date ______________

Skills Worksheet

Vocabulary and Section Summary

Mirrors and Lenses

VOCABULARY

In your own words, write a definition of the following terms in the space provided.

1. plane mirror

____

____

2. concave mirror

____

____

3. convex mirror

____

____

4. lens

____

____

5. convex lens

____

____

6. concave lens

____

____

SECTION SUMMARY

Read the following section summary.

- Rays are arrows that show the path of a single light wave.
- Ray diagrams can be used to find where images are formed by mirrors and lenses.
- Plane mirrors and convex mirrors produce virtual images. Concave mirrors produce both real images and virtual images.
- Convex lenses produce both real images and virtual images. Concave lenses produce only virtual images.

Name ______________________ Class ______________ Date ______________

Skills Worksheet

Vocabulary and Section Summary

Light and Sight

VOCABULARY

In your own words, write a definition of the following terms in the space provided.

1. nearsightedness

____

____

2. farsightedness

____

____

SECTION SUMMARY

Read the following section summary.

- The human eye has several parts, including the cornea, the pupil, the iris, the lens, and the retina.
- Nearsightedness and farsightedness happen when light is not focused on the retina. Both problems can be corrected with glasses or eye surgery.
- Color deficiency is a condition in which cones in the retina respond to the wrong colors.
- Eye surgery can correct some vision problems.

Name ______________________ Class ______________ Date ____________

Skills Worksheet

Vocabulary and Section Summary

Light and Technology

VOCABULARY

In your own words, write a definition of the following terms in the space provided.

1. laser

____

____

2. hologram

____

____

SECTION SUMMARY

Read the following section summary.

- Optical instruments, such as cameras, telescopes, and microscopes, are devices that help people make observations.
- Lasers are devices that produce intense, coherent light of only one wavelength and color. Lasers produce light by a process called *stimulated emission.*
- Optical fibers transmit light over long distances.
- Polarized light contains light waves that vibrate in only one direction.
- Cordless telephones are a combination of a telephone and a radio. Information is transmitted in the form of radio waves between the handset and the base.
- Cellular phones transmit information in the form of microwaves to and from antennas.
- Satellite television is broadcast by microwaves from satellites in space.
- GPS is a navigation system that uses microwave signals sent by a network of satellites in space.